The New Technology Elite

How Great Companies Optimize Both Technology Consumption and Production

Vinnie Mirchandani

John Wiley & Sons, Inc.

Published by John Wiley & Sons, Inc., Hoboken, New Jersey.
Published simultaneously in Canada.

For general information on our other products and services or for technical support, please contact our Customer Care Department within the United States at (800) 762-2974, outside the United States at (317) 572-3993 or fax (317) 572-4002.

Wiley also publishes its books in a variety of electronic formats. Some content that appears in print may not be available in electronic books. For more information about Wiley products, visit our web site at www.wiley.com.

Library of Congress Cataloging-in-Publication Data:

Mirchandani, Vinnie.
The new technology elite : how great companies optimize both technology consumption and production / Vinnie Mirchandani.
pages cm
Includes index.
ISBN 978-1-118-10313-5 (hardback); ISBN 978-1-118-22390-1 (ebk); ISBN 978-1-118-23727-4 (ebk); ISBN 978-1-118-26217-7 (ebk)
1. Technological innovations–Management. 2. Organizational change. 3. Industrial management. I. Title.
HD45.M536 2012
658–dc23

2011048578

Printed in the United States of America

10 9 8 7 6 5 4 3 2 1

To Margaret, Rita, Tommy, and Peanuts—the second opportunity to spend time with you has been so special.

Contents

Preface

"This is a song of hope."

That's how the band Led Zeppelin led off many of its live performances of what many consider the best ever rock-and-roll hit, "Stairway to Heaven."

Well, this is a book of hope.

I started my technology career in the 1980s when there was palpable excitement about technology providing strategic advantage. Then technology, IT in most companies, went into the woodshed for the next two decades. It focused on costs, controls, and compliance. It was not focused on competitive advantage. In fact, its costs and overruns made many companies uncompetitive. Over the past 15 years, I have helped countless clients focus on cutting IT costs. That's not much fun.

Starting several years ago, that two-decades-old hope flickered again, and as I wrote my last book, *The New Polymath*, I saw an amazing amount of technology-enabled innovation being planned.

This book builds on that hope. It comes from cataloging elite technology athletes I have included in the book and how they plan to improve how we live, work, and play.

How this book came together deserves an explanation.

If you have ever seen a Gartner (the technology research firm) presentation, you know it is the antithesis of the old-school McKinsey (the strategy consulting firm) presentation, which dictated "no more than six bullets per slide, six words per bullet." Gartner slides have dense graphics, the handouts have speaker notes in a small font, and the voiceover allows the speaker a chance to present additional perspectives beyond the graphs and the notes. You also get to speak uninterrupted at their events for 45 minutes before any questions. My five years there have influenced my presentation style ever since.

So imagine my challenge at the Ignite conference in Toronto, the night of September 2, 2010.[1] The format allowed each speaker just five minutes to present 20 slides. No exceptions. So I had five minutes to summarize my theme—the 400 pages in the book, *The New Polymath.* The setting was even more challenging. It was at the historic Drake Hotel, but in a comedy club format. The audience, many in their twenties, stood in the dark sipping their drinks—the comedy part came from watching the speaker struggle as the projector relentlessly moved slides every 15 seconds.

Through the 20 slides, I tried to convey one basic message to the young crowd. There was something very wrong when the Twitter stream of the GE Global Research Center profiled in the book (and named EdisonsDesk after its famous founder, Thomas Edison) had only 704 followers up to 3,500 as of December 2011 when Britney Spears had almost 6 million. The young are enamored (as I see with my own teenagers) with their iPhones, their Facebook friends, and Twitter streams. They should also be aware that there is plenty of "compound innovation" going on at GE, BASF, BMW, Hospira, and other case studies in the *Polymath* book that are blending a wide range of infotech, biotech, cleantech, healthtech, and nanotech.

On the flight back home on that trip, I found myself thinking about my presentation at Ignite. Interestingly, it was a much younger me standing in the audience asking two questions. Okay, so I would benefit from learning more about GE and BMW and Hospira, but flip it around—what are those big companies doing to learn about and develop products for the younger, tech-savvy consumers like me and others in the audience? And if Twitter allows Britney, Ashton and a bunch of other popular folks to have millions of followers, just how massive is the Twitter technology infrastructure and that of Facebook and Google?

Those two questions became the seed for this book. They led me to focus on them in the research for my innovation blog[2] and in conversations with a number of my consulting clients and industry colleagues.

At the Consumer Electronics Show in Las Vegas in early 2011, I got a clear answer to the first question. While the big excitement during the show was around the tens of new tablets expected to be rolled out later in the year, I observed companies from just about every industry—from Walgreens, the pharmacy chain, to Nike, the shoe company, to Ford, the auto company—were showing off technology-enabled "smart" products for the new tech-savvy consumer.

It is a really exciting time for many of these companies. For too long IT has been an expensive and low-payback back-office investment. Now technology in the companies' products is allowing them to generate revenue and growth. Technology is fun and profitable again.

The answers to the second question came in the Apple disclosure that it had sold 25 million iPads in its first year, that Google had signed over 20 million users for its + service in its first month, and that Facebook had crossed the 750-million-user threshold (that is more than the population of most countries in the world). The more I analyzed the operations of Apple, Google, Facebook, Amazon, Twitter, and eBay—their data centers, distribution centers, retail stores, application ecosystems, global supply chains—the more I was impressed with the "industrialization" of their technology. They are considered "consumer" tech, but they have better technology on a greater scale than most enterprises.

Traditional technology users are embedding technology in their "smart" products and services thus learning to become technology vendors. Technology vendors like Apple are, in reverse, running retail operations better than Nordstrom. eBay's PayPal unit is running better operations than many banks. Amazon is running logistics better than many distributors. Google is running data centers far more efficiently than IBM's or EDS's. They are the new best practice leaders.

It hit me that the traditional distinction between technology user and vendor is outdated. The baseball term "switch-hitter" came to mind. Not just in baseball—in soccer, athletes who can fire rockets with either leg are valued. Same with many other sports—ambidexterity is a much sought-after attribute. Similarly you have to learn to be comfortable on both sides of the plate, to be ambidextrous in the development and consumption of technology.

Beyond ambidexterity, though, truly elite athletes contribute in other ways. Baseball switch-hitters are more valuable if they are also good fielders, base stealers, and have strong, accurate throwing arms. The elite are multi-dimensional—they play good defense and offense.

This book is about elite companies—the technology version of Gold Glove fielders and stolen base leaders. And the great news is the more I researched, the more I found many of them. Across industries. Across countries.

In this book I have tried to bring out that diversity. There are more than 100 examples and interviews from New Zealand to South Korea, from farming to municipal services. The 17 case studies and four guest columns bring the elite attributes out in detail.

Flow of the Book

The book is organized in three parts.

Part I sets the stage for the technology "buyor"—companies that are becoming efficient buyers *and* vendors of technology. It also explores the landscape for such behavior across varied industries and geographies.

Part II explores 12 attributes of becoming what I call the technology elite. It is no longer about being able to talk geeky terms like HTML5 or SQL Injection or cloud architectures. It is about product design elegance, about physical presence in strategic retail locations, about ecosystems of developers and thriving App Stores. It is about being paranoid in the world of groups like LulzSec and Anonymous. It is about being pragmatic in a world where attorneys are even more influential than engineers. It is about being able to fly to Xiamen or Xanadu at a moment's notice.

Part III is focused on how regulators, Wall Street and other market analysts, and society in general are learning to cope with the tsunami of technology innovation that is headed our way. The technology elite we catalog are being sensitive to these external influences—and preparing for the world where these influencers themselves become more tech-savvy.

Let's go through each part and chapter individually.

Part I: The Convergence of Technology Production and Consumption

Chapter 1: The New Monday Morning Quarterback. Traditionally classified as technology "buyers" and "user organizations," many companies are learning to embed technology in their products and themselves become technology "vendors." They are learning to sell to the new tech-savvy consumer who is ready for new form/factors in every product, including shirts, pens, cars, and bulbs. Other companies like 3M and GE would be offended if you mentioned they were "learning" to become technology vendors. They have long viewed themselves as technology vendors even if Silicon Valley may disagree. Then there are companies like UPS, which calls itself "about half a transportation company, half a technology company." The case study on UPS details its technology innovations in DIADs, its franchise stores, its aviation innovations, and much more.

Chapter 2: The "Industrialization" of Technology. Apple, Amazon, Google, eBay, Facebook, Zynga, and Twitter are mostly focused on consumer-facing technology. Under the covers, however, they show the power of "industrialization of technology" as they scale to massive numbers—25 million iPads in Year 1 and 20 million users for Google+ in its first month of introduction. These are technology vendors emulating best practices from varied industries, even competing with them and taking them to new levels. They tend to focus on each other. Amazon, Google, and Apple are all vying for streaming music service. Amazon competes with eBay for online commerce, Google competes with Twitter, Facebook, and Zynga for social impact. Google competes with Apple and PayPal/eBay for mobile dominance. In the process, though, they are raising the bar for established technology vendors like IBM and Microsoft and for corporate IT. The case study in this chapter focuses on HP's complex global supply chain and how it adjusts to countless hiccups caused by the Japanese tsunami, the Iceland volcano, and also to longer-term shifts.

Chapter 3: Amazon to Zipcar: No Industry Untouched. Either out of paranoia, promise, or "phoenix thinking," industry after industry is thinking about next-generation, technology-influenced products and services. It is almost impossible to find an industry with what Warren Buffett, the legendary investor, calls "unbreachable moats," where technology's influence is not felt. The case study associated with this chapter is tiny Roosevelt Island in the East River in New York and its amazingly ambitious vision of leveraging technology like smart parking and a next-generation ferry service.

Chapter 4: Australia to Zanzibar: No Country for Old Products. In some ways we live in a flat world where Apple, Google, and Facebook have become our common language. On the other hand, there are so many unique nuances that it would be naïve to assume global branding and universal product versions can be successful for all products. The different rates of technological evolution also offer significant policy opportunities as states, provinces, and nations compete for jobs and investments. The case study for this chapter focuses on the small Baltic country of Estonia and its remarkable digital "Tiger Leap" and rapid evolution away from decades of communist stagnation.

Chapter 5: Convergence, Crossover, and Beyond. Chapter 1 showed traditional technology consumers who are learning to become technology vendors. In Chapter 2, we saw technology vendors who are delivering scale and best practices from a variety of technology buyer industries. Just as in baseball, where only 15 percent of players are switch-hitters, the ambidextrous technologist who is good at both technology production and technology consumption is an elusive species. We need them in the form of "crossover" executives who bring experiences from both technology vendor and user organizations. We present guest column perspectives of two such crossover executives: Tony Scott, CIO at Microsoft who got there after stints at Disney and GM, and Vijay Ravindran, Chief Digital Officer at the Washington Post Co., who was at Amazon in its formative stages.

Part II: Key Attributes for the New Technology Elite: Three Es, Three Ms, Three Ps, and Three Ss

Chapter 6: Elegant: In a World of Flashing 12s. We are seeing a revolution in design in devices, in our software, and in our architecture. If you aspire to be one of the technology elite, you have to put industrial design high on your agenda. In the case study associated with this chapter, we see how Virgin America, using good design and technology, is redefining the airline industry, an industry that in some studies scores lower in customer satisfaction than the IRS.

Chapter 7: Exponential: Leveraging Ecosystems. A thick application catalog has always been important for a technology platform's success going back to IBM's huge success with the AS/400 in the 1980s. It's become dramatically more important in the past few years. In particular, Apple and Google have shown their App Stores can scale to hundreds of thousands of applications and billions of downloads. In the case study for this chapter, we analyze how RIM (the BlackBerry company) has had to react to this new world.

Chapter 8: Efficient: Amid Massive Technology Waste. There is massive waste in technology: printer ink at $5,000 a gallon, roaming charges of $4 a minute from some countries, calls to application contact centers that amortize to over $10,000 each. The technology elite don't just focus on innovation to improve the top line; they are also intensely focused on efficiencies. In this chapter's case study, we outline the countless efficiencies Facebook has delivered in its Prineville data center which opened in 2011. In a technology world that is traditionally secretive, Facebook surprisingly shared details of much of the data center as part of its Open Compute Initiative.

Chapter 9: Mobile: If It's Tuesday, It Must Be Xiamen. To become a tech elite you have to be a Marco Polo and a Gulliver—bravely exploring a fast-changing world. Suppliers and captive units in exotic locations can provide unique competitive advantage. They can also make much more complicated the supply chain and product development cycles. The case study here shows how Boeing learned from a global supply chain for its new 787. The experience was painful with many delays, but also delivered an amazing number of innovations. The case study further describes how Boeing used HCL Technologies as a glue to bond many of those widespread elements.

Chapter 10: Maverick: No Rules. Just Right. In technology, more than any sector, being a maverick is tolerated, even encouraged. Disruption is not a dirty word in technology. Tech elites understand, though, being a maverick does not mean no rules. It means defining your own rules and discipline to support your position. The case study here is Apple, which has shown time and time again how it breaks others' rules, while defining new ones for itself and its competitors. We look at 10 gutsy moves Apple has made over the last decade and how it plays Maverick—the character Tom Cruise played in the movie *Top Gun*.

Chapter 11: Malleable: Business Model Innovations. Technology-elite companies like Apple and Amazon have shown that creativity in pricing and efficiency in costing are as important as good product design and logistics. And not just pricing around their own products but around music, books, and telephone service. Technology is allowing every industry to experiment with new business models like "as a service," and "from the bottom of the pyramid." In the case study, we see how a young company, Valence Health, using sophisticated analytical technology and a new business model aimed at providers versus payers of healthcare, is building a viable venture.

Chapter 12: Physical: Why Test Driving Is Still Important Even in a Digital World. We would not buy a car without a test drive and somehow we seem to have forgotten that physical, tactile experience continues to be important with technology products. The technology elite know better and also know you need knowledgeable, friendly customer service to go with it. The case study for this chapter is Taubman Shopping Centers, which has thrived in the last decade with Apple, Bose, and other technology stores and its own digital infrastructure, when brick and mortar was supposed to be dying.

Chapter 13: Paranoid: But Not Paralyzed. Teardowns, jailbreaks, and rootings are almost a badge of honor in technology world. They are tame, however, compared to the malicious hacking and espionage technology companies are increasingly subjected to. In such a climate, it helps to be paranoid. The case study here profiles the Wireless Aerial Surveillance Platform, a drone that can hack into networks and eavesdrop on mobile conversations and shows other reasons to be vigilant. Of course, being paranoid does not mean being paralyzed. The technology elite just look at it as a cost of doing business. Life has to go on.

Chapter 14: Pragmatic: When Attorneys Influence Technology Even More than Engineers. If the hacking and the espionage described in the previous chapter do not paralyze companies looking to build smart products, the spreading lawsuits surely can. In many ways, the technology elite know that good lawyers are just as important as engineers in technology. They can help enterprises to be pragmatic, even when surrounded by "rattlesnakes." Benjamin Kern, an attorney with the firm of McGuire-Woods, also has an interesting personal background as a technology entrepreneur. He provides, in a guest column, some of that pragmatic advice on the tricky world of fuzzy patents and uneven intellectual property (IP) protection with global supply chains.

Chapter 15: Speedy: In a New Era of Perishability. A key trait of the new technology elite is their speed—in product innovation, in anticipating changes in competitive landscapes, in managing volatility in demand forecasting and supply chains, and even in their back office. In the case study we look at the Corning Gorilla Glass product, which defines the new clockspeed. It has been adopted in over 400 electronic products in less than three years.

Chapter 16: Social: Amid Chatty Humans and Things. The technology elite know how we live in a world of chatter, human and nonhuman. Learning to interpret that chatter and to magnify it via social savvy is no longer a "nice to have." Not just your employees, your products also need to be "social." In the case study we look at a socially savvy product—the Lexmark Genesis—and the social media launch it enjoyed.

Chapter 17: Sustainable: Mining the Green Gold. To be considered a technology elite, it is increasingly expected that you put sustainability high on your self-evaluation scorecard. The definition of sustainability, however, gets more ambitious by the day as we grapple with "conflict minerals" and the ethics of becoming dependent on nonrenewable rare earths in so many of our technology products. The case study looks at Google's breathtaking array of green initiatives.

Part III: Outside Influences on the Technology Elite

Chapter 18: Regulators and Technology. Many of the earlier chapters show savvy government and municipal technology groups. In general, however, regulation of technology lags the fast-changing markets it is supposed to watch over. Regulators are themselves being challenged to become technologically elite. The case study focuses on how 3M uses its "Periodic Table" to summarize its 46 technology platforms (that get manifested in over 55,000 of its products) to communicate with market watchers. It manifests what regulators will increasingly have to become savvy about.

Chapter 19: Society's Changing View of Technology. While we can build technologically elite enterprises, we cannot mandate a technologically elite society. The reality is it is "unelite" and uneven. So we need a new set of professionals to prepare society for the avalanche of coming technologies. In a guest column, Prof. Mary Cronin of Boston College focuses on another challenge we will face in the next few years. She writes about the continuous, automatic, and invisible tracking of individuals by multiple smart devices and the related explosion of personal data and the new privacy challenges society will have to address.

Chapter 20: Market Watchers Morphing. As the technology elite like Amazon, 3M, HP, and UPS pioneer new ways of communicating to financial and other market analysts, they raise the bar for those analysts. They also create new expectations for every other company they are benchmarked against. We present the text of a groundbreaking letter Jeff Bezos, CEO of Amazon, sent to his shareholders, which is full of technology jargon. It does come with this reassurance: "Now, if the eyes of some shareowners dutifully reading this letter are by this point glazing over, I will awaken you by pointing out that, in my opinion, these techniques are not idly pursued—they lead directly to free cash flow."

And finally, we end with a summary chapter that brings together many of the trends we have discussed in the first 20.

Endgame: "Welcome to the NFL." This chapter summarizes the 12 elite attributes we discussed in Part II. It also profiles the impact of more demanding regulators and market watchers and changing societal expectations we discussed in Part III.

What Got Left Out?

I could have profiled twice as many guest columns and detailed case studies. I could have written about all kinds of advances in biotech and nanotech since my last book. What about peers and competitors of companies profiled in this book—are they sitting still or innovating on their own?

It's like making a Hollywood movie. Plenty gets left behind in the "director's cut." In my case, some of those "30 minutes" edited out can be found on my two blogs. Those are living, breathing documents compared to a printed book that can be only a snapshot.

Of course, I would welcome reader comments and conversations. I expect a few will disagree with me, for including the HP supply chain example when it announced, then reversed course within a matter of weeks, that it was de-emphasizing its PC business. In my opinion it is a good example of the acrobatics needed in the dynamic technology marketplace. Others might disagree with my profiling RIM with its work-in-progress ecosystem as it tries to catch up to Apple's and Google's. Or the much-delayed Boeing 787. In my opinion, the 787 supply chain and the passenger comfort innovations it incorporates deserve the ink. So, I look forward to the feedback and the discussions.

In the meantime, as "Stairway" advises, there's time to change the road you are on. It's a good time to emulate the technology elite. Yes, this is a book of that hope.

Acknowledgments

As I reviewed a late edit of this book, I felt like I was watching ESPN's X Games, which focuses on extreme sports. The pages that follow are filled with many technology athletes and their acrobatics. So, let me start by thanking the "athletes" themselves. The many interviewees and guest columnists in the book spent hours talking to me, wrote about unique nuances in their businesses and provided their innovative perspectives on technology.

The comparison to the Games also reminded me that hundreds of coordinators have a hand, behind the scenes, in organizing such an event.

There were many like Barry Dayton at 3M, Bryan Majewski at Baker Tilly, Sarah Pakyala at Corning, Alan Alper at Cognizant, Alison Bolen at SAS, Tiffany Anderson at Tibco, Ramana Rao at The Washington Post Company, and Lacey Higgins at Workday, who coordinated many of the interviews in the book. Jim Spath at Stanley Black & Decker helped with a critical review of the materials. They may not show up in the book, but these and many others played a significant role. To them, also my heartfelt thanks.

It's the same with my publisher, John Wiley & Sons. John DeRemigis, and his entire editing and marketing team, worked marvels through several versions of the manuscript. A book on technology, business, society, and public policy stretches the dictionary in many directions.

The biggest thanks go to my wife, Margaret. She encouraged me to write the book. I would have waited a few more years after *The New Polymath*. She calmly helped me navigate the ups and downs common in such an undertaking.

She is the unheralded Chief Organizer of the event. Or her behalf, and all the "athletes," I invite you to enjoy the "Games"!

Part I

The Convergence of Technology Production and Consumption

Chapter 1

The New Monday Morning Quarterback

For a week every January, Las Vegas replaces Silicon Valley as the technology hub of the world. The International Consumer Electronics Show (CES) attracts 150,000 exhibitors, attendees, government officials, investors, and fans from around the world.[1] They come to see an orgy of technologies at the show that even outshines Vegas's other outlandish attractions.

The show has traditionally been a harbinger of technology and societal trends with the products that are launched there: 1981 saw the introduction of the camcorder, 1991 the Interactive CD, 2001 the Microsoft Xbox. Each year in between and after there have been other spectacular announcements.

The year 2011 was no different. It will likely go down as the "Year of the Tablet." Over 100 options from Motorola, Dell, Samsung, Toshiba, and others competed for attention at CES.[2] They were all hoping to match the phenomenal launch of the Apple iPad a few months prior.

If the 7,000 journalists, bloggers, and analysts at the show were not exhausted from analyzing the varied tablet form/factors and new features, they were chasing down rumors at the show about whether Apple was about to launch a Verizon version of the iPhone. (The rumor was later confirmed as Apple provided an option to the AT&T network that was previously an exclusive for the iPhone in the U.S. market.)

Lost in the excitement about iPad killers and iPhone rumors at the show was an even more significant nugget—the list of exhibitors included companies from just about every non-technology vertical industry.

There was Walgreens—yes, the pharmacy chain—showing off its Refill application that allows you to scan the bar code from a previous prescription using a mobile phone, transmit it, and get a text message to go pick it up at a nearby store. At many of its stores, you could use drive-through lanes and related technologies.

Whirlpool showcased its Duet washer/dryers with LCD screens and various laundry apps designed to give users advice on stain removal and other laundry questions.

Nike introduced a GPS-enabled Sportwatch developed in collaboration with the navigation vendor Tom Tom.

Ingersoll Rand showed off tech innovations around its Schlage home security and Trane thermostat products.

Ford chose to unveil its all-electric Focus at the CES show rather than at the traditional car launch showplace, the Detroit Auto Show, which was only a week later. In a later guest column for *Fortune,* Bill Ford, Executive Chairman, wrote, "Many of Ford's suppliers are now nontraditional suppliers like Microsoft and retailers such as Best Buy, which are helping provide the charging and IT infrastructure for this new form of mobility."[3]

Not to be outdone by Ford, GM showcased a retail, boxed version of an OnStar-equipped rearview mirror. This opened up OnStar to nearly any vehicle—from a Ford to a Toyota. Its features such as automatic crash response, turn-by-turn navigation, stolen vehicle location assistance, emergency and roadside services, and hands-free calling were long a reason to buy a GM car with the built-in OnStar.[4]

3M showed off its Patterned Transparent Conductors (PTC). Using technology that enables a high degree of pattern control of conductive materials on flexible substrates, 3M is able to produce conductive traces down to two microns wide or less to support projected capacitive touch sensing. This capability supports the development of new touch-enabled consumer electronic devices. PTC is able to decrease the amount of space needed on the device bezel. Using silver, it offers a significantly lowered resistance that allows sensors to support fast response times, even in tablet sizes.

What's going on here? These companies live far from Silicon Valley and are known as retailers and auto companies. Why are they competing for booth space and geek attention with technology vendors?

The Monday Morning Letdown

What's going on, in Malcolm Frank's words, is the "Sunday Night versus Monday Morning phenomenon." Frank, SVP of Strategy and Marketing at Cognizant, is regaling an audience of CIOs at a customer conference in Orlando. The slide shows a picture of *Rumors*—the Fleetwood Mac 1977 album. His audience giggles as his voiceover tells the cynicism of his young sons as they happened upon his record collection and he tried to justify why we bought complete albums back then, not just individual MP3 tunes. The reaction of his sons only accentuates what he is seeing at work.

At Cognizant and in other outsourcing firms he has previously worked at, Frank has always been surrounded by young workers—a theme common in technology services. He sees their Monday morning reaction—no Facebook listing of colleagues? No iPads? No thumb drives allowed? This is pretty striking after Sunday night at home with HDTV, Flip video cameras, Microsoft Kinects, Bose audio gear, and Apple retina displays.

Yes, it is a trend called "consumerization of technology," and Apple gets plenty of credit for it.

> Of course, it was not just Apple. Microsoft introduced the Xbox in 2001; HP announced more of a focus on entertainment with its Media Center PC in 2002; Skype arrived in 2003 and would change calling habits for millions around the world; the massively multiplayer game

World of Warcraft disrupted our lives in 2004; and in 2005, with the Sony Rootkit issue, consumers started to hear about intellectual property issues previously limited to corporate corridors. Since 2005, Google's growing Web and mobile presence has introduced consumers to analytical power that big companies paid bucketloads of money for. The Apple iPhone—considered one of the most successful product launches ever—and all kinds of GPS, gaming, entertainment, and other gadgets have only accelerated the consumerization trend. The Wii reshaped our expectations of computer interfaces; the Kindle, our expectations of books. *Electronic House* magazine has celebrated houses with all kinds of home theaters and elaborate security systems. eBay has built a cottage industry of individuals doing business from home. JetBlue has moved call reservations to agents working from home. SOHO is no longer just a Manhattan neighborhood but a growing revenue category for technology companies as the acronym for small office, home office.[5]

TAF—The Technologically Advanced Family

Matt Murphy and Mary Meeker of the venture capital firm Kleiner Perkins brought out the global consumer technology proliferation numbers vividly in a presentation they gave in February 2011.[6] They include 18.8 trillion mobile minutes, 972 million Google users, 10 billion Apple applications downloaded, and 130 million active Zynga users.

We are just getting started. Globally, only 14 percent of the world is on a 3G or faster network, so the market for smart phones and mobile applications is still embryonic.

Then there is the potential demographic stretch. Mark Zuckerberg, CEO of Facebook, thinks the age requirement of 13 is too restrictive for users of his social network. He says, "My philosophy is that for education you need to start at a really, really young age."[7] He does have a point. Many families get even younger children mobile phones, encouraged by billing plans that cost as little as $9.99 a month for the extra line. Nintendo has a workaround for kids under 13. Their policy states, "For children under the age of 13, we have developed a Family Account system. In order for a child under 13 to have a Club Nintendo account, the child's parent or guardian must register for Club Nintendo and create a Family Account. That parent or guardian can then create sub-accounts for other family members."[8]

On the other end of the age spectrum, companies like General Electric and Intel are beginning to market a new type of patient care called "aging in place." These services rely on new monitoring technology that may effectively replace retirement homes and assisted living centers.[9] Instead of being intimidated by e-book readers like the Kindle, many elderly readers say they often find them lighter to hold than books, can increase font size, and use the "text to speech" feature to have the book read to them.

In the middle we have digital minimalists like Kelly Sutton, the founder of CultofLess.com, a website that has helped him sell or give away most of his possessions. He was left with his laptop, an iPad, an Amazon Kindle, two external hard drives, a "few" articles of clothing, and bedsheets for a mattress.

Sutton got rid of much of his clutter because he felt the ever-increasing number of available digital goods have provided adequate replacements for his former physical possessions.

"I think cutting down on physical commodities in general might be a trend of my generation—cutting down on physical commodities that can be replaced by digital counterparts will be a fact," said Mr. Sutton."[10]

Then there is the phenomenon of the "technologically advanced family."

Joshua Moore is 12 years old, born and raised in Silicon Valley, and the child of two software industry executives. His house has always been filled with digital devices, making Joshua very comfortable being immersed in technology. His baby pictures were taken on a digital camera, transmitted to his relatives via webcam, and shared on Ofoto.com as soon as that site came online in 1999.

Josh has three computers, a tablet, an Android and WebOS phone, an iPod Touch, an Xbox Kinect, a Wii, and a slew of other electronic devices on his desk in his bedroom. He uses these devices to produce a popular YouTube channel and for his blog on GottaBeMobile.com. His digital camera has recorded videos and taken pictures from the top of Half Dome in Yosemite and Mount Haleakala in Maui, and 40 feet down while scuba diving. His parents don't pay him any allowance, because he earns money on his own with his blog, as a Google YouTube Partner, and with Google AdSense.

Josh is just one of the digitally advanced members of his family. His older sister, Adriana, seriously considered only sending invitations to her upcoming wedding via Facebook and is always connected to her friends on her iPhone. His mother, Young, used her MacBook Pro to study for and take the California Bar exam. And his father, Dennis, is a veteran of established software companies like Oracle and SAP, as well as start-ups (including C3 Energy, which has plenty of buzz even though it has been in stealth mode for years), and is a popular blogger himself. Dennis even started one of the first "corporate alumni" groups—The OracAlumni Network—to help friends and colleagues from Oracle to stay in touch and professionally network. The whole family stays in synch via Google Calendar, which allows them to keep their busy schedules open enough to have time for family brunch or dinner and to make sure that Josh always has a ride. After all, it's not all about bits and bytes.

Maybe it's all in the DNA. Dennis learned to program at age 13 at the New York City public library and then at NYU's Courant Institute. He got a TRS-80 in 1977, and worked his way through Princeton as a programmer before moving out to Silicon Valley in the late 1980s. Dennis and Young met when they were both working at Oracle Corporation, where Young was then one of the few female executives. Three of Dennis's five siblings are also in the computer industry, in roles ranging from digital advertising to programming.

Josh thinks it's quaint when his parents describe things like darkrooms, telephones with dials, TVs without remotes, going to the bank, waiting until after 7 p.m. to make an affordable phone call, having to go to the library to look something up, and screens you can't manipulate by touch. Of course, it is a fair bet his own kids will think that today's tablets and cell phones are antiquated.

BYOT—Bring Your Own Technology

BYOT is an acronym increasingly heard in corporate IT circles. It is the recognition that with so much consumer technology, enterprises could be leveraging it, not fighting it.

Unisys ran a poll that asked, "What percentage of the cost of your job's IT tools would you be willing to fund if you had freedom to choose what you could use?" Nearly one-third (32 percent) of the respondents said that they would be willing to pick up the full cost. Twenty-one percent said that they would pay up to half of the cost, and another 21 percent said that they would fund up to 30 percent of the cost.[11]

Savvy CEOs have been walking in to their CIOs and asking, "Where's my iPad?" And no, not offering to pay 30 percent of the cost!

The lightbulb has gone on in even savvier CEOs. "If the consumer lives in the world of iPads and Kindles, can our own product be rethought to be more appealing to this tech-savvy consumer?"

New Form/Factors in Every Product

Welcome to the new Monday morning quarterback. This consumer is willing to try new technology-enabled form/factors in just about everything.

- There is the "smart" pen and pad from LiveScribe. The Echo records everything you doodle on its special dot paper pad and then allows you to plug into your computer and see the exact replica in digital format. It also has a voice recorder so you could be recording that interview or conversation while you are taking notes.
- There is the smart shirt for athletes from Under Armour. *Wired* magazine describes the UA E39 as "It weighs less than 4.5 ounce and is made from the same material as the rest of the company's line of compression-based apparel. Yet just below the sternum, the shirt also contains a removable sensor pack called a 'bug' that holds a triaxial accelerometer, a processor, and two gigabytes of storage. The information collected can be broadcast via Bluetooth to smartphones, iPads, and laptops so that scouts and trainers can view the power and efficiency of each athlete's movements."[12]
- PixelOptics offers the smart version of bifocal or progressive eyeglasses. It alters the focal power of the lens when you tilt your head or tap the frame.

- The Plaza in New York has developed the smart hotel room. The iPad in each room allows you to order room service, make restaurant reservations, book wake-up calls, print out boarding passes, and control the room's lighting and air conditioning.
- There is the smart restaurant—Do (as in Dough) in Atlanta, Georgia. Not only are paper menus replaced by iPads, the tablets can also be used to tell the valet to pull up your car. They also allow patrons to change music while they marvel in the brilliant video graphics that cover every wall of the restaurant. The tech-enhanced bathrooms boast motion-sensored hand dryers and sinks with iPad "mirrors" positioned on the walls.[13]
- Moen is pitching its smart shower. The IOdigital wall-mounted control panel, with a handheld remote, lets you set and maintain water temperatures. For baths, it also allows you to set the water level. And these are pretty unimpressive compared to the features of the vertical spas the company offers!
- There is the smart lawnmower courtesy of Toro. The TimeCutter SS brings speed controls to Zero Turn technology Toro has pioneered. Choosing the "high mode" is meant for maximum speed for mowing the flat, open spaces in the yard. The "low mode" allows for enhanced maneuverability when cutting around trees, landscaping, or other tight spaces.
- Rain Bird has for decades made lawn maintenance "smart" by allowing us to control irrigation times for various zones. Its ESP-SMT irrigation controller makes that much smarter. It utilizes historical and real-time weather data (you input data like your zip code, allowed watering days, and the plant/soil type for each zone) to determine optimum watering needs of the landscape based on the on-site current weather conditions. It calculates the site's evapotranspiration (ET) rate each day and then deducts effective rainfall amounts from the ET rate to determine how much water is needed to maintain the optimum level of moisture in the soil.
- There is the smart version of the long indispensable level from Stanley. Once the user has downloaded the application, he or she begins by calibrating the level with a gravity-driven (bubble) level to create a simulation that takes advantage of the accelerometer in the mobile device.

Not Just "Smarter" Products, also "Smarter" Services

Not just products, but also look at how services are becoming smarter.

- USAA Bank supports mobile deposits. Use the camera on your phone to take an image of the back and front of the check and voilà—no need to go to a branch or even to an ATM machine to make a deposit. It is also partnering with PayPal to allow customers to pay almost anyone with an email address or mobile phone number.
- Progressive Insurance offers Snapshot, a small telematics device that connects to the insured car's electronic diagnostic port. It allows Progressive to analyze data from the device on your driving patterns and if safe, it promises up to 30 percent discount on your premium.
- State Farm Insurance has In-Drive, which "mashes" up the GM OnStar FMV and Progressive's Snapshot just mentioned. Offered in collaboration with Hughes Telematics, a small device allows drivers to access emergency roadside assistance, vehicle diagnostics, and maintenance reminders. And if you allow State Farm to collect data like vehicle mileage, speed, and braking performance, it offers insurance discounts up to 50 percent.[14]
- There's Nationwide Insurance's iPhone application. It allows you to collect and exchange accident information, take pictures of the accident scene, start a claim, display your insurance card and other account information, and locate nearby tow trucks and authorized repair shops.
- There's EPB Fiber Optics (Chattanooga's municipally owned fiber-to-the-premises network), which offers 1 GBPS broadband to consumers—that's 100 times faster than what the average telephone or cable company offers residents in most other states.
- The Hamilton County, Indiana, sheriff's office has a next-gen 911 (emergency) call center. Each agent in the office can view "five large screens [that] simultaneously show call status, caller information, police radio activity and other data—all of which can be shared instantly over radio, phone, Internet, dispatch and cellular systems."[15]
- There's the SmartMeter from utilities like PG&E that allows consumers to monitor their hourly energy usage and better manage their electric bills.

- Over the past 15 years, starting with a deployment in Texas, there has been a steady rollout across the United States and broadly across North America of the AMBER (America's Missing: Broadcast Emergency Response) alert system, which is triggered when young kids are abducted (the program honors Amber Hagerman, who was abducted when she was nine). In 2011, the U.S. FBI released a Child ID app for the iPhone. It allows parents to store on their iPhones recent pictures of their children along with vital statistics—age, height, weight, distinguishing marks—to easily and quickly share with law enforcement in case a child goes missing.

These are available to the average consumer. When it comes to professional and corporate customers, the version of smart is much smarter. In fact, they are not products, but "platforms" that allow companies to sell a variety of services, accessories, and ancillary products.

- Take the Emerson platform, Trellis, for data center management. It is a portfolio of hardware, software, and services designed to allow data center managers to optimize efficiency, availability, and capacity utilization.
- John Deere and its FarmSight initiative, which is planned to help farmers in three areas: Machine Optimization for increased up-time, Logistics Optimization for better fleet management, and Decision Support with user-friendly monitors, sensors, and wireless networks to enable access to machinery and agronomic data.
- Take Siemens, the German giant, which offers a variety of telematic services with transponder, satellite, and other technology to cities around the world as they roll out electronic toll collections in their version of smart services to their citizens.
- Georgia-Pacific (G-P) has its enMotion, which automatically dispenses towels at the wave of a hand. It's perfect for restaurant and other commercial washrooms and kitchen sinks. The dispensers are hygienic, they lower waste with predefined sheet lengths, and can be customized with time delays and sensor ranges. More importantly, it allows G-P to sell various types of paper (its prime product), air fresheners, and other products.

So, the Genie Grants You Your Smart Product Wish

Back to the CEO and the CIO conversation. Why are our products not similarly smart?

So, you meet the proverbial Genie and he grants you your wish. Your product is smart now. But as the story goes—are you sure you want that wish granted? Are you ready for your new world? You are now a Technology Vendor, and that means:

- *Getting used to technology product half-lives.* A sign of the times is Best Buy, which sells plenty of extended warranties with mobile devices, TVs, and digital cameras. It has introduced its version of "shortened" warranties via its Buy Back program. As customers lust after newer versions of everything, that program promises (for an upfront fee) to buy back anytime within two years of an electronics purchase date (four years for TVs), giving you a gift card to use toward another purchase.
- *Adjusting to Moore's Law.* Not every industry adjusts easily to price/performance improvements common in technology. Take auto dealers. Initial versions of DVD players, navigation systems, and Bluetooth speakers all sold for thousands of dollars. Dealers continue to expect those prices even as handheld Garmin GPS units and BlueAnt speakerphones are available at a fraction of the cost. Importantly, these devices are portable, so customers can take them when they travel or switch from one family car to another. Indeed, there is a cottage industry of FM transmitters, power inverters, backup cameras, portable satellite radios, coolers, and other gadgets you can buy for your car, many for less than $20.
- *Rethinking product documentation.* In a world where no training is needed to navigate Amazon.com and any "live" help is a keystroke and online chat away, companies still ship products with clumsy, printed product manuals.
- *Understanding technology law.* Technology licensing and other legal nuances require a different mindset. As an example, Nike+, which allows users to track information collected by sensors in its shoes such as the elapsed time of the workout, the distance traveled, pace, or calories burned, has four pages of terms of use on its website.

- *Getting used to competition from left field*. Whirlpool has over 40 percent of the U.S. appliance market and historically has looked at companies like GE and Sears as competition. Soon, though, as appliances get smarter and the guts become more software and sensors, competition will also come from Korean companies like LG and Samsung, which have traditionally been focused on electronic products. Young-ha Lee, president and CEO of LG's Home Appliance Company, said at the CES event described earlier, "This year will be the beginning of a new era of home appliances. By that I mean that we have reached the tipping point where appliances are now run entirely by CPUs and computer code. Just as automobiles became rolling computers a decade ago, home appliances are experiencing the same transformation."[16]

If Siemens, a company founded in 1847, Toro, founded in 1914, and Moen, founded in 1937, can do it, so can anybody else. They are not startups by a long shot.

Of course there is the nightmare scenario. If our products are not viewed as smart, what about the risk from standing still? Will our customers increasingly view them as dumb?

Conclusion

Traditionally classified as technology "buyers" and "user organizations," many companies are learning to embed technology in their products and themselves become technology "vendors." Others, like 3M and GE, would be offended if you mentioned they were "learning" to become technology vendors. They have long viewed themselves as technology vendors even if Silicon Valley may disagree. Then there are companies like UPS, which calls itself "about half a transportation company, half a technology company."

Let's look at UPS in detail in the case study that follows.

Case Study: UPS—That's Technology "Amore"

There will be no more stress, 'cause you've called UPS ... THAT'S LOGISTICS.

In the fall of 2010, UPS launched a global TV ad campaign around the Dean Martin classic "That's Amore," sung in Mandarin, Spanish, and English. The lyrics went along with the theme of "We (Heart) Logistics," which replaced UPS's previous theme of "What can Brown do for you?"

UPS is unabashedly proud of its "brown collar," "box kicker, label licker," "logistics is sexy" image. Its 2010 revenues of almost $50 billion and a history of over a century justify that pride.

Look under its zillions of bar-coded labels, however, and you see a marvel of technology that keeps a massive data center in Mahwah, New Jersey, and another outside Atlanta, Georgia, humming. Twelve mainframes rated at 52,000 MIPS, nearly 16,000 servers, and 190,000 workstations are part of its technology landscape.[17] UPS says its package tracking is done by one of the world's largest DB2 site (the IBM database software). It stores over 10 petabytes of data. It is one of the largest users of mobile minutes in the world. UPS.com, available in 32 languages, handles 48 million tracking requests on a peak day during holiday season. Its technology helps optimize truck routes and famously minimize left turns. It is investing in a variety of fuel-efficient trucks and techniques, such as continuous descent approach, to glide its planes for fuel efficiency and noise reduction.

Tom Ryan, at TKR Consulting Associates, is a former analyst at Gartner and Aberdeen Group (the technology research firms) who specializes in logistics, supply chain, and transportation solutions. He adds, "UPS has been one of the leaders in leveraging technology for competitive advantage in the marketplace, especially computer and automatic identification technology. In its early days as a small package delivery company, UPS employed multiple 'large forehead' type people to build route optimization software for all of their delivery routes worldwide. They then designed and had built custom mobile terminals to allow their drivers to be more efficient in their route tasks as well as provide real-time updates to their corporate tracking systems. In order to eliminate manual typing errors by their operations staff, UPS leveraged existing bar-code technology and invented one of the first two-dimensional bar codes (PDF417), so that one scan of one bar code could convey all the data about the shipment that was needed. UPS used this real-time data to build one of the leading in-transit visibility systems so that from pickup to delivery each customer can know where their stuff was 'when it had wheels on.' As UPS expanded its transportation logistics services from small packages to less-than-truckload and more, it brought this optimized highly visible infrastructure to play."

Scott Davis, the Chairman and CEO of UPS, agrees with Ryan's assessment, when he says, "Often when I'm talking to investors I tell them that we're about half a transportation company, half a technology company. And 20 years ago that wouldn't have been the case. Twenty years ago we were a package transportation company."[18]

Let's look at four of countless aspects of this "half a technology company." Let's start with the ubiquitous Mr. Brown—the UPS driver most of us are familiar with.

The DIAD

A Delivery Information Acquisition Device (DIAD) is what UPS's truck drivers use to guide their delivery routes and more. The DIADs help UPS drivers in their brown trucks deliver over 22 million packages a day during the peak holiday season. It was introduced back in 1990, way before the iPhone or the iPad was even conceived of. And its battery lasts all day—much longer than most Apple or Android devices.

Jackie Woods of UPS says:

> The DIAD has been a solid foundation for UPS to not only provide outstanding service to our customers, but also has been an invaluable tool for our drivers to improve day-to-day efficiencies for more than 20 years. What started as a device to provide customers with proof of delivery visibility and to collect timecard information from our drivers has grown into a personal assistant of sorts providing a wealth

of detailed information on package deliveries and pickups, including customer preferences and specific requests, while the UPS driver is on-road. The device also uses GPS to provide guidance to our drivers to ensure packages are delivered to the right location and to maintain our standard of service excellence.

The capabilities of the device have grown significantly over the years, moving from the first generation device, which did not have the ability to communicate, to the fifth generation, which incorporates an industry-first for rugged, handheld devices—the ability to communicate over both CDMA and GPRS cellular technologies within the same device. This will allow UPS to increase our coverage, experience fewer outages and leverage best-cost carrier options.

Strengthening our coverage footprint is essential in continuing to service "day of" customer requests for pickup or delivery changes to help manage customers' changing business or personal needs. The DIAD V has additional features such as a camera, dimensional imaging, enhanced wireless LAN capabilities, touch screen and keypad, more memory and faster CPU, and is approximately half the size and weight of its predecessor, the currently deployed DIAD IV. The DIAD V began deployment to test sites in June of 2011. Deployment of more than 100,000 devices to more than 70 countries continues through 2013.

Wood continues:

The journey through the five generations of DIAD devices has been an interesting one. Up through the fourth generation device, UPS has influenced the rugged handheld industry in significant ways, including defining requirements for device communications (LAN, WAN, Bluetooth), imaging capabilities, and durability requirements. As you can imagine, a device used by UPS drivers has to support features that allow us to best service our customers and withstand harsh environmental factors such as weather, wide temperature ranges, and drops to concrete—all while presenting a professional image to our customers.

UPS has made a significant non-recurring engineering (NRE) investment over the years with hardware vendors (Honeywell is replacing Symbol [part of Motorola], which provided earlier generations) to achieve the required functionality and durability to run our business. The most recent generation, DIAD V, has allowed us to realize the fruits of our labor as it is the first DIAD we were able to purchase

> "off the shelf" with limited customizations (like the brown color, of course. Previous customization required country and regulatory certifications for the device and accessories for each country in which the equipment would be used so there are other efficiencies to leverage).
>
> We continue to work closely with industry partners to help define and influence their product roadmaps. However, at this point, with the rapid evolution of smart phones and hand held devices, our hope is that the industry advancements will outpace our needs and moving new devices into UPS operations will be as common place as replacing a cell phone.

An electronic manifest is downloaded into DIADs as drivers start their workday, so the driver can see each scheduled package delivery displayed on the DIAD in the exact order of delivery needed. If a driver is about to deliver a package to the wrong customer, or is forgetting to deliver one of a group of packages to a customer, the DIAD alerts the driver with an audible alarm. UPS tries to minimize changes to the day's schedule because unlike other carriers, a UPS driver is trained to deliver multiple services—pick up and deliver overnight packages, collect COD payments, and pick up and deliver ground packages to the commercial and residential customers on his or her route. Urgent customer pick-up messages can, however, be transmitted in color-code to the DIAD to alert the driver.

UPS says the DIAD allows it to eliminate the use of about 90 million sheets of paper per year, sparing over 7,000 trees.

The Green Initiatives

It makes sense for a company with a huge fleet of planes, trucks, and workstations to look for fuel and energy efficiencies. Some of the industry-leading innovations UPS has pioneered include the following.

Continuous Descent Approach

A UPS cargo airline pilot lines up with the runway much earlier—sometimes up to 40 miles away from landing—and cuts the thrust once and descends at a consistent rate.

UPS pioneered the approach and found nitrogen oxide emissions dropped by 34 percent below 3,000 feet and engine noise fell by 30 percent within 15 miles of the airport. The planes also saved 250 to 465 pounds of fuel per flight.[19]

Package Flow Technology

CIO David Barnes (who happens to also be one of the most respected CIOs in the industry) started his career at UPS as a part-time loader over 30 years ago and describes the then and now:

> If you were a pre-loader in the morning, it was a great college part-time job. Get in there before class. But you have to memorize the routes of three drivers. A very painful process because if you misloaded that package and you put 101 Main St. behind something else, the driver had to loop back and that's not very efficient. . . .We'd done it one way for 90-plus years at the time. We are very good at it—industry leading. How could you do it dramatically different? We threw out all the norms and developed a system called Package Flow Technology. Just one small aspect of that program was to take a fresh look at how you optimize networks. And as a result of optimization routines that we put within that software within the practices at UPS, we were able to reduce our mileage in a given year in the United States alone by 30 million miles—that's about 3 million gallons of fuel. And perhaps more important to all of us it represents 32 million tons less of CO_2.

Smart Pickup

This new service, introduced in 2010, is designed for customers who want the convenience of a scheduled pickup but who may not ship a package every day. The service ensures that a UPS driver stops at a customer location to pick up a package only when necessary. The service is expected to eliminate 8 million miles from the total driven by UPS each year in the United States and will save an estimated 793,000 gallons of fuel and 7,800 metric tonnes of CO_2 emissions.

Telematics Analysis

In this initiative UPS monitored the driver's—and the truck's—workday. It captured data on more than 200 vehicle-related elements from speed, RPMs, and oil pressure to seatbelt use, the number of times the truck is placed in reverse, and the amount of time spent idling. The data was uploaded when a driver returned to his or her center at the end of the day, to the data center in Mahwah, New Jersey. At the two Georgia test sites, the initiative helped slash the amount of time trucks idled by 24 minutes per driver per day—a fuel savings of $188 per driver per year. Multiply that by more than 90,000 U.S. package drivers, and the potential savings is significant. It also helped reduce maintenance costs. Currently, the company replaces a starter approximately every two years, whether a vehicle makes 30 or 150 stops a day. The telematics allows UPS mechanics to base the decision instead on things like the actual cycles of each starter, and the amount of voltage it draws when it's used.

Data Center Efficiencies

CIO Barnes again:

> In the Atlanta (data center) we chill the cold water that runs through those chillers at night. We have a 500-gallon tank buried beneath the data center. We chill the water at night. We get lower electricity rates at night and then we use that chilled water during the hot summer days to chill the plant down. It's very, very environmentally sensitive. It reduces the demand for power generation at peak times during summer days.

And:

> I think when most of us started years ago (the data center) was always a refrigerated room when you walked in there. It was humidity control for good purposes and, by the way, still good purposes—but very cold. Today if you walked in ours they are lukewarm. And so the issue of having to chill these things down really isn't something that's very efficient. So we are running them a lot hotter.

The UPS Store

Scott Galloway cringes when you ask him about his 15 years in the ERP software industry. He was a marketing executive with Powercerv and then president of Verticent, a spin-off software company focused on the metals industry.

"'The long sales cycles, the long implementation projects, the low customer ROI—and we were far less complex than SAP, Oracle and some of the other larger software vendors," he says.

He traded that career for an investment in a UPS Store franchise. It is Store # 4,586 in Tampa, Florida. Galloway says, "This is simplicity personified compared to the career in ERP." Simplicity sounds like an oxymoron when you are also talking about the largest DB2 database in world. Few of the over 4,000 store operators in the United States have a technology background like Galloway does. So, UPS strives to keep the processes and technologies simplified and standardized.

In 2001, UPS acquired the Mail Boxes Etc. franchise. Rebranded as The UPS Store, it has steadily grown the volume of stores and the range of products in such stores. Besides the mainstay shipping, the stores offer copying and printing services, packaging supplies, mailboxes, notary services, and plenty more. The shipping, of course, allows for a wide range of global destinations via multiple air, ground, and other paths. It has to account for millions of rates, corporate discounts, and taxes.

The store deals primarily with four software modules—the Customer Manifest System (CMS), the Mailbox Manager, the Point of Sale (POS), and Quickbooks (Administration) from Intuit.

The CMS is constantly being updated by UPS for changes in rates, taxes, insurance, customs requirements, and for new products like digital printing—which is increasingly emailed in and customers stop by to just pick up the finished job. The Mailbox Manager is updated for integration with postal rules, as in form 1583-A to authorize them to deliver mail in your post box at the store. The Point of Sale is updated to add ever more scanning and messaging formats so package delivery status or mailbox renewal information can be emailed to customers. Quickbooks is the standard version that you and I can buy. The chart of accounts is uniquely configured for UPS store franchises. And in an extremely useful and powerful service, UPS aggregates and shares regular financial information sent by franchises and other shipping trends from its mainframes. You can see the huge leverage each store gets from UPS's scale when it comes to such benchmark data and market trend intelligence.

More standardization beyond the software comes into play when you look at the equipment in the store. The scanner gun to read bar codes is from Symbol (the Motorola unit). The weighing scale is from Mettler-Toledo and the label printer is from Epson.

Such standardization also facilitates training. Galloway raves both about the range of web courses (hundreds of them in short, five-minute modules ideal for staff training) and the in-person training available. Galloway runs the store part-time with his wife and one staff member. As part of his induction into the franchise, he spent a week with another store operator, did a two-week course at the franchise headquarters in San Diego (UPS corporate is in Atlanta), and then another in-person week in another store.

While many of the operators are curious, they are not expected to keep up with how UPS planes are innovating with continuous descent approaches or how their delivery trucks minimize left turns or what UPS is planning in their fifth-generation DIADs as we described earlier.

The Industry Solutions

UPS has increasingly been using its vast logistics network to provide industry solutions.

As its website says, "Especially over the last decade, we've added a lot of capabilities that many people would never associate with UPS, such as repairing laptops for a computer manufacturer, filling prescriptions for medical devices, and providing online printing services."

In the automotive industry, it has worked with GM around the supply chain for warranty parts; in healthcare, it has helped French pharmaceutical company Boiron with warehouse management; in high-tech, it has helped Cisco build out its European supply chain; and in government, it has worked with several agencies at various national and regional levels.

Each industry brings its own set of compliance and business nuances. UPS has hundreds of customers in the industries mentioned above and others.

What If UPS Were a "Full Technology Company"?

What if UPS were not as the CEO described "half a transportation company, half a technology company"? Could it be a full-fledged technology company?

Think of the DIADs going back to 1990 and their huge leverage with the mountain of mobile minutes they procure—UPS could have introduced an iPhone-like device way ahead of Apple. Think of the green innovations—UPS could easily be a cleantech player. Think of the store software—UPS could easily be a software vendor to countless other industries that use franchise models. Think of the various verticals UPS has built niches in.

That's idle thinking. UPS is clearly comfortable in its skin as a dominant logistics provider and where technology fits in that definition. As its jingle goes:

Where technology knows right where everything goes . . . THAT'S LOGISTICS. . . . Bells will ring and ding and ding and ding . . . THAT'S LOGISTICS.

Chapter 2

The "Industrialization" of Technology

Kevin O'Marah lives and breathes logistics. A senior analyst at Gartner, the research firm, he has focused for years on the science and magic behind logistics. His team analyzes practices and challenges in a wide range of industries including aerospace, auto, consumer products, and life sciences.

Each year the supply chain team at Gartner publishes a list of companies they respect across industry lines. For the past four years (2008–2011), Apple has topped that list.[1]

The Apple—Physical and Digital—Supply Chain Mastery

O'Marah says,

> Apple's leadership has been pervasive. Traditional supply chain disciplines like managing an extended network of contract manufacturers

> and component suppliers are fully in force, but beyond these areas Apple has led in at least two vital ways. First is in its huge leverage of the digital supply chain. By fostering the development of a secondary market in applications for its iPhone, the company has shown again (as with iTunes) that consumer product revenue growth with zero inventory is not only possible, but repeatable. The other area in which Apple's supply chain leadership is increasingly important is in the retail experience. As one of a handful of deeply vertically integrated brands, Apple's retail chain achieves almost unimaginable success in its stores.

While Apple gets plenty of kudos for the elegance of its products, peel the onion, and, as we will show in various chapters, it has built a retail store chain that is the envy of even long-time retailers. It has built an elaborate global network of suppliers and contract manufacturers that has confused the traditional accounting that economists use to measure global trade. In addition to the elaborate physical supply chain, it has had to integrate the digital supply chain as iPhones are activated via iTunes at customer homes and via carriers. As it rolls out its iCloud, it has built one of the biggest data centers in the world. It has built an ecosystem of apps and games around its products at a scale never seen before. Admirably, it built its supply chain in a much more volatile industry than that of consumer products or chemicals.

Of course, Apple has itself driven the high-tech industry volatility with its own pace of product introductions. Dell used to be held as a benchmark of efficiency with its "build-to-order" supply chain. It manufactured mostly what was ordered and even paid for in advance. Apple raised the bar by showcasing a new product, guesstimating likely demand, and tuning its supply chain day by day and hour by hour. It broke traditional rules of demand forecasting—because there was little historical data to forecast from for a version 1.0 iPod or iPhone or iPad. It balanced the risk of overproducing or increasing buffer inventory and taking write-offs versus underproducing and losing customers to the next competitive product just a few weeks away. It took that risk, time and again, and made the rest of the industry do the same. And the risks are not small when you are talking 3 million iPads in the first quarter of introduction.[2] The Dell model has been "out-Delled." As a Dell executive acknowledged, the old model worked well for 20 years but the "environment has changed."[3] Dell, no slouch, is reshaping itself to this new world where we are not building to order but to anticipated demand.

Retail wizardry, contract manufacturing, supply chain efficiencies—we used to look to companies like Walmart, Fedex, and Procter & Gamble for leadership in these practices. Now we are seeing well-run technology vendors like Apple become the "best practice" paragons.

Not Just Apple

Not everyone agrees with the Gartner supply chain rankings. Some think its methodology weighs financial performance too heavily. Others think it does not factor sustainability enough.

An Apple competitor thinks Gartner is being too generous to Apple and other technology vendors in its ranking. "I would not call them or us a textbook supply chain management leader. Having said that, the elite who run operations in high-tech are all darned good at execution and adjustments." He says that as his team furiously reworks supplier scenarios impacted by the Japanese tsunami in early 2011. "Not to be unkind about the tragedy, but this is what many of us live for—the adrenalin rush from such disruptions."

As we will see in the HP case study later in this chapter, volcanoes, explosions at supplier factories, labor issues at suppliers, and shipping sea lane delays are fairly common disruptions in the technology supply chain.

"Constant course correction" is how one former executive refers to the approach. "If the executive team decides to change direction, it's instantaneous," this ex-Apple honcho says. "Everybody thinks it's a grand strategy. It's not."[4]

Transitioning from "make-to-order" to "make-to-constant-course-correction" is tricky, to say the least. Zynga, the social gaming company famous for Farmville and popularity with Facebook users, launched CityVille in 2010. "IT staff watched participation on CityVille build slowly the first few days—at that pace, it might have taken two years to reach the first million participants. Then registration started to ramp up. Four months after launch, CityVille had brought 20 million new users to Zynga's daily computing workload."[5]

In early 2011, a document circulated about Amazon.com that had language like:

> The diversity of products demands that we employ modern regression techniques like trained random forests of decision trees to flexibly incorporate thousands of product attributes at rank time.

What was remarkable about the document was that it was not a discussion aimed at university math majors or even attendees at a technology conference. It was part of the annual shareholder letter Jeff Bezos, CEO of Amazon.com, sent to his investors. (The complete letter is shown in Chapter 20.)

The letter went on to include statements like:

> Service-oriented architecture—or SOA—is the fundamental building abstraction for Amazon technologies. Thanks to a thoughtful and far-sighted team of engineers and architects, this approach was applied at Amazon long before SOA became a buzzword in the industry.[6]

The financials that followed highlighted that Amazon does not just have large distribution centers around the world. It was well on track to sell over $1 billion in 2011 in services to companies like Zynga above (Zynga uses Amazon's elastic cloud when it launches products and brings the processing in-house when a game hits a more predictable customer level) from massive data centers it has set up around the world. Amazon's e-reader Kindle and e-books are expected to generate over $5 billion in revenues in 2011.[7]

Amazon is a legitimate technology vendor feared and respected by its competitors. Indeed, its subsidiary Lab126, which designed the Kindle, is based in Cupertino, California, a stone's throw from Apple's headquarters. What distinguishes Amazon even more is its savvy with physical logistics, which was tested from the beginning by competition with Walmart and other retailers. As it has consistently used discounted shipping as a competitive advantage, it has been sophisticated in its design and location of distribution centers. It has innovated with "postal injections," in which it uses its own trucks or independent carriers to drive truckloads of orders to local postal depots from Amazon warehouses. The procedure eliminates processing steps for the U.S. Postal Service.[8]

Google is known for its search and mobile technologies, but even more impressive are its investments and prowess in clean energy innovation. For its data center and other operations, Google buys renewable energy from various wind farms and has invested in large solar panel farms and in geothermal companies. It has won approval from the Federal Energy Regulatory Commission (FERC) to become an electric utility. At a data center event it hosted in Zurich, Google shared what it has learned from operating efficient data centers, "laying out the best practices that any IT manager can do on a reasonable budget, such as maintaining hot and cool aisles, managing airflow, measuring the energy consumption in the facility, using outside air whenever available, running the data center at a higher temperature like 80 degrees, and optimizing power distribution."[9] We further discuss Google's green initiatives in Chapter 17.

Even More Examples—eBay, Facebook, and Twitter

The PayPal unit of eBay, in business slightly over a decade, has revolutionized the online and mobile payment world for millions (particularly individuals and small businesses) in 190 countries. In 2010, it processed over $90 billion in payments.[10] To do this, it has implemented technology to handle the intricacies of regulatory controls across countries, protection against fraud, tracking money laundering, and more.

As mobile payments gain traction, this technology is even more in the crosshairs of much more established banks, credit card companies, other technology vendors, and even giant telecom companies. Telecom companies hope to take advantage of next-generation mobile phones with near-field communications (NFC) so you can pay by tapping the phone on a specialized reader at the store.

eBay has also demonstrated remarkable growth in its mobile marketplace business. CEO John Donahue told analysts in January 2011, "We extended our leadership position in mobile commerce, nearly tripling our eBay mobile (general merchandise value or GMV) year over year to nearly $2 billion, with strong holiday shopping momentum. In 2011, we expect mobile GMV to double to $4 billion. Mobile is clearly becoming a new way people shop. Our mobile apps have been downloaded more than 30 million times in eight languages across 190 countries."

In 2011, Facebook opened the doors to its large data center in Prineville, Oregon, which at full capacity will be 300,000 square feet. Facebook needs such large centers to handle its explosive growth of over 750 million active users and over 3 billion photos posted by users each month. What was particularly interesting was the design of the center:

> It has spent two years developing everything that goes inside its data centers—from the servers to the battery cabinets to back up the servers—to be as green and cheap as possible. For example, Facebook uses fewer batteries thanks to its designs, and to illustrate how integrated the whole computer operation is, the house fans and the fans on the servers are coupled together. Motion-sensitive LED lighting is also used inside. The result is a data center with a power usage effectiveness ratio of 1.07. That compares to an EPA-defined industry best practice of 1.5, and 1.5 in Facebook's leased facilities.[11]

We will discuss this data center in more detail in Chapter 8.

While critics say it is mostly frivolous and there are numerous jokes about its frequent "fail whale" when the site goes down, it is remarkable the influence Twitter has had across the world. That is especially true when you consider it is just a five-year-old company with a staff of fewer than 500. Whether it be a catalog of the uprising in Egypt or the Japanese tsunami, or even an inadvertent live tweet during the Osama bin Laden raid, as Brian Solis has observed, "The new reality of a real-time world is that news no longer breaks, it Tweets."[12]

What is fascinating is the simplicity behind the design of the service. As the *MIT Technology Review* wrote in 2007, "(Co-founder) Stone says the company completed a working version of the software in only two weeks using Ruby on Rails, a programming language and a set of prefabricated software modules widely employed by developers of the new raft of Web services known as Web 2.0. The hard part, he says, was 'navigating the business aspects of the mobile industry. It took us months to get a short code and figure out how to play nice with all the major U.S. and international mobile carriers.' "[13]

While it may have started simply, Twitter has scaled immensely, as it described its new search architecture in 2010: "With over 1,000 TPS (Tweets/sec) and 12,000 QPS (queries/sec) = over 1 billion queries per day (!) we already put a very high load on our machines. As we want the new system to last for several years, the goal was to support at least an order of magnitude more load."[14]

Conclusion

Apple, Amazon, Google, eBay, Facebook, Twitter, and Zynga are mostly focused on consumer-facing technology, but under the covers show the power of "industrialization of technology." These are technology vendors emulating best practices from varied industries, even competing with them and taking them to new levels. They tend to focus on each other. Amazon, Google, and Apple are all vying for streaming music service.[15] Amazon competes with eBay for online commerce and so on. In the process, they are raising the bar for established technology vendors like IBM and Microsoft and for corporate IT. In Chapter 1 we saw how technology "buyers" were becoming vendors. The vendors described in this chapter and HP with its complex global chain described in the following case study are behaving more like traditional technology buyers—setting new benchmarks of productivity and efficiency.

Case Study: HP—The Quest for a "10 Out of 10" Supply Chain

"It could be your darkest hour or your finest," says Tony Prophet as he discusses the impact of the Iceland volcano explosion, the Japanese tsunami, and other disruptions on HP's supply chain. Prophet, a senior vice president, oversees hardware purchasing and logistics for HP.

It is a massive operation, the largest in the technology industry, with over $60 billion in components, warehouse, transportation, and other logistics costs. The HP machine churns out two personal computers a second, two printers a second, and a data center server every 15 seconds.[16] It is also a constantly evolving operation with a changing mix of company-owned factories and contract manufacturing utilizing air/ship/rail logistics from/to most countries around the world. So those challenging "hours" Prophet talks about come with alarming regularity.

In August 2011, HP announced it was looking at strategic options around its PC business; as the market moves to tablets and mobile devices, and as the company sought to invest more in software markets. Under a new CEO, Meg Whitman, it reversed course a few weeks later, citing, among other things, the impressive supply chain performance of the hardware business we describe.

The Short-Term Acrobatics

One of those defining hours Prophet talks about was after the Eyjafjallajokull volcano explosion in Iceland in April 2010. That event ended up disrupting flights to northern European airports for weeks. HP flies many of its products from Shanghai, China, to those airports. On many days, HP is the biggest single buyer of outbound airfreight from Shanghai. Fully loaded 747-400 cargo planes with HP notebooks and other products often take off three times a day from that airport.

Prophet's team immediately decided to switch shipments destined for northern airports like Frankfurt, Germany, to southern airports like Barcelona, Spain, and Naples, Italy. The problem was they could not find sufficient commercial flights for those destinations from China. So they booked charter flights. Even as HP products were arriving in Europe, competitors found charter freight prices had gone up 50 percent or were not available anymore. HP's 72-hour lead turned out to be significant.

In March 2011, Prophet was woken up at 3:30 a.m. to be informed of the massive Japanese earthquake and tsunami.

The *New York Times* wrote, "Japan is the world's third-largest economy, and a vital supplier of parts and equipment for major industries like computers, electronics, and automobiles. The worst of the damage was northeast of Tokyo, near the quake's epicenter, though Japan's manufacturing heartland is farther south. But greater problems will emerge if rolling electrical blackouts and transportation disruptions across the country continue for long."[17]

Within two hours, HP had set up a crisis management team. Japanese component manufacturers are a key part of HP and competitors' products, so the HP team had to quickly assess the status of component inventory and its safety. Twenty-four hours later, the second risk-management phase began—to acquire additional buffer inventories of components in case the crisis caused a shortage or increase in prices. HP is the largest purchaser in the industry of many of the components like memory, so hedging against the disruption was critical.

"There are so many other examples of disruptions I could list," continues Prophet.

There are explosions at supplier factories, labor issues at suppliers, and shipping sea lane delays. When you run a complex, global supply chain, there are thousands and thousands of links that can be disrupted and require rapid remediation.

Prophet chuckles when he talks about other spikes that happen on a regular basis. "We obviously work with our channel partners to plan for increased demand around back-to-school campaigns, Black Friday promotions, or regional holidays like the Chinese New Year. Those are significant events to plan for, but compared to the havoc from the volcano or the tsunami, those are so much easier to manage."

The Long-Term Shifts

If those short-term maneuvers are dazzling to watch, even more impressive are the long-term shifts Prophet talks about.

> One of our ongoing projects has been to streamline our supply chain, especially after major acquisitions like Compaq or Palm or 3Com.

The HP-Compaq merger, a particularly large one, afforded plenty of streamlining opportunities over the past several years. "In 2006, we had a combination of 70 (HP and Compaq) supply chain nodes worldwide, most company-owned and in relatively high-cost locations. Also, most were geared to making desktop PCs and the notebook market was taking off."

As Prophet told a meeting of financial analysts in December 2010:[18]

> Each of those nodes had inbound hubs, outbound hubs, unique IT connections, and drove a tremendous amount of overhead to support the complexity of these nodes.

Now it has been streamlined to

> About 30 nodes and significantly fewer of them company-owned. Obviously biased more towards lower-cost locations, but not exclusively. We continue to operate a plant in Indianapolis. We continue to have a plant in Japan, so where it makes sense for proximity to the customers, to serve those customers with high velocity, we're there.

Prophet continues:

> That's just the PC business. Now layer on our imaging, printing, enterprise server, network nodes. At one point we had more than 450 nodes across all our supply-chain networks. So just in Southern California, we had three warehouses within 30 miles of each other. So printing equipment, PCs, and notebooks would come on different shipments and go to different warehouses—and get shipped separately to the same distribution centers, as many were often going to the same customer. The warehouses, the containers, the trucks were not always fully utilized. So there were plenty of opportunities to consolidate real estate, shipping, and trucking and optimize all those costs.

> Another opportunity was to move to standardized components across product lines. So, not to buy connectors all over the place. We have moved from about 60 percent standardized and centrally procured three years ago to over 85 percent.
>
> That has brought another advantage. A few years ago, we bought many components and commodities on the spot market. We have shifted to long-term commitments to key suppliers. The suppliers can make longer-term capacity investments and we don't have to be as dependent on the vagaries of the volatile market.

So, there are times when HP overpays compared to the spot market. Can you imagine, though, being exposed to the spot market for critical Japanese components after the tsunami?

Finally, there are process and IT opportunities. As Prophet told that meeting of analysts, "Within the Personal Systems Group, the PC business, we have 65 ways of planning, You've got one for notebook and one for desktop, one for consumer notebook and one for commercial notebook, one in EMEA and one in Latin America, et cetera, et cetera."

Across the whole business, HP had more than 1,000 processes and 300 IT applications. "So our objective is to cut the processes by an order of magnitude and cut the IT applications by more than a third. We're moving to common IT applications to drive those processes, and so that to our suppliers and to our customers we look like one company."

"We've got a path, a strategy charted to build what we think will be a '10 out of 10' supply chain, and we're about a year into this transformation."

You cannot serialize these projects. They all have to march forward in parallel and they all have to adjust to the short-term hiccups and shifts in the market as the recent market move in many Western markets to tablets away from notebooks.

Another fascinating dimension of the HP supply chain is its global reach and its pioneering investment in Western China.

The Global Dimension

Prophet, in that presentation to analysts, highlighted some of HP's global reach.

> First multinational to manufacture PCs in Russia.
>
> Upgraded our operations in north central India.
>
> Significantly upgraded our operations in Brazil.
>
> We believe we'll be the first multinational manufacturing PCs in Turkey. That'll allow us to effectively serve central Asia, Eastern Europe, and the Middle East, the Mediterranean region particularly.

Particularly impressive is the HP pioneering expansion into Western China.

> China's west contains 70 percent of its land mass, 30 percent of its population, but just 20 percent of its total economic output. Per capita GDP in the mega-city of Chongqing, which has a population of 27 million, was USD 3,500 in 2009, versus USD 10,500 in Shanghai.[19]

It has been called the "New Wild West" and there have been multiple attempts to develop that part of the country that have been hampered by: "poor infrastructure, minute levels of outside investment, an ailing environment (especially in Chengdu, Sichuan Province, and Chongqing Municipality), and a weaker education system than the rest of China."[20]

Prophet continues:

> We have been in China over two decades, and like other Western companies we could see the spiraling real estate costs, employee absenteeism, stretched infrastructure in the developed East and South. Starting in 2007 we started to evaluate alternatives like Vietnam and Malaysia. But the more we looked at Western China, the more we also saw the opportunity for the domestic market. If you draw a circle out 750 miles from Chongqing, you are looking at about 300 million people—which by itself makes it one of the largest PC and other gadget markets in the world. So that is our first focus here in the west of the country—made in China, for China.

Indeed, the Chongqing plant has already produced for the domestic market what is being called a "rural" laptop—designed to handle intense heat and rain—and that model is also seeing demand in other regional countries.

Even more impressive is the infrastructure the Chinese are building around Chongqing. There is the high-speed freight rail line to the port city of Shenzhen.[21] The route run by a contractor, Cosco, takes roughly 50 hours and emphasizes "five fixed" services—fixed stops, fixed trains, fixed timetables, fixed routes, and fixed rates. Shenzhen is well equipped to handle bulk shipping to most major destination markets.

Next, with encouragement from HP, the airport at Chongqing was extended by 400 meters in a rapid construction project to allow fully laden 747s (with merchandise and extra fuel) to fly nonstop to Europe.[22]

An even more exciting development is that of the next "Orient Express"—a rail service between Chongqing and Duisburg, Germany, covering nearly 7,000 miles in 13 days. That is 26 days quicker than the current rail-sea combination, and considerably cheaper than the air option. And it should get even quicker as China shares its growing high-speed rail experience with countries along the path—Kazakhstan, Russia, and Poland.

Summarizes Prophet, "We are going to see significant competitive advantage from our investment in Western China. Chongqing is expected to produce 20 percent of the world's laptops and we are proud to be one of the pioneers here."

Prophet says he is even prouder of the leadership role HP has taken with sustainability initiatives.

Sustainability

Prophet provides a number of examples:

> We took a leadership role in forming the EICC (Electronic Industry Citizenship Coalition), which promotes an industry code of conduct for global electronics supply chains to improve working and environmental conditions.[23]
>
> We have been transparent in naming our top 100 suppliers and disclosing the carbon footprint of our extended supply chain.
>
> We have set goals beyond those required for any jurisdiction, including reducing BFR/PVC[24] and accountability for conflict minerals.[25]

We discuss sustainability in the tech sector further in Chapter 16.

"A Great Operating Engine"

Ray Lane, Chairman of HP (and a partner at the venture capital firm Kleiner Perkins), says about Prophet, "I first met Tony as a rising star in the supply chain practice of Booz Allen (now Booz and Co.) in the '80s. He was part of the brilliant team of industrial engineers and other geniuses that my partner, Dan Lewis, put together in our Cleveland office, which was then a mecca for logistical leadership. That early disciplined grooming continues to show in the amazing results Tony has consistently delivered for the HP supply chain over the last several years."

Prophet is more modest—he says he is proud to be part of HP, which he calls "a great operating engine." As he told the analysts in the presentation discussed previously, "This wasn't a single individual driving this. We structurally changed things. We put rigorous processes in place to make these things happen. So it wasn't a personality or an individual driving it. It's a company driving it, 300,000 people strong."

At the Consumer Electronics Show in Las Vegas in January 2011, the dominant device was a tablet and HP seemed like a distant runner-up to Apple. The same show a year later was dominated by Ultrabooks, and HP looked impressive again, and smart for not having spun off its PC division.

Chapter 3

From Amazon to Zipcar: No Industry Untouched

A survey from the National Association of Corporate Directors and the consulting firm of Oliver Wyman provides a sobering assessment of the tech-savviness of most boards of directors:

> While some management teams have kept pace with rapidly changing technologies to succeed in today's business environment, it is the very rare board that has been able to provide the governance and leadership that is so desperately needed in this area.[1]

While you could argue whether they are tech-savvy, most boards are certainly paranoid about being "Blockbustered."

The *New Yorker* summarizes well the company's missteps: "Blockbuster's huge investment, both literally and psychologically, in traditional stores made it slow to recognize the Web's importance: in 2002, it was still calling the Net a 'niche' market. And it wasn't just the Net. Blockbuster was late on everything—online rentals, Redbox-style kiosks, streaming video."[2]

No wonder executives flocked to events like IBM's "Smarter Industries Symposium" in Barcelona, Spain, in 2010. More than 1,200 clients, partners, and IBM leaders from around the world gathered together to discuss how organizations could harness the enormous potential from "smarter" technologies, solutions, and business processes. Out of that event, IBM generated reports with ideas for 10 industry sectors—Banking, Communications, Electronics, Automotive, & Aerospace, Energy & Utilities, Government, Healthcare, Insurance, Oil & Gas, Retail, and Transportation.[3]

Technology-Driven Turmoil

Why stop at just those 10 industries? History has shown technology-driven turmoil in a much wider set of markets:

Businesses that can be digitized: Monster.com and other job sites decimated newspaper help-wanted ad revenues. U.S. newspaper help-wanted ad revenues declined 90 percent from 2001 to 2010.[4] Craigslist and Yahoo! took away other classified revenues from the newspapers. iTunes, Walmart.com, and other MP3 sites killed record stores and CD sales. Now, streaming music options threaten MP3 players. eTrade and other online brokerage firms took sizable chunks of trading revenues from old-line firms. Pandora and other customized web-based radio stations are threatening FM radio.

Labor-intensive industries: Call centers and various knowledge worker industries have seen business move to lower-cost labor markets as global telecommunications improved. In reverse, tools like Google Translate and voice recognition technologies have impacted translation and transcription services. Travelocity, Orbitz, and other reservation sites took plenty of business away from travel agents. Intuit prodded H&R Block to become a software vendor while diversifying away from its tax preparation services. QR codes on home "for sale" signs that allow users to look up websites with home walk-through videos are reducing the role of the traditional real estate agent.

Real estate intensive industries: Netflix, Redbox, and others tormented Blockbuster in video rentals, as we saw earlier. Amazon, even as it partnered with Borders in book sales, showed the weakness of its brick-and-mortar dependence. Best Buy is evaluating smaller stores, airport kiosks, and other lighter real estate footprint options. Small and home office options have challenged many a downtown real estate landscape.

Other asset-intensive industries: Zipcar is changing the rental car industry by using technology to do away with expensive parking structures, rental offices, and large fleets. Telepresence rooms from Cisco and robots from VGo are starting to challenge air travel. Home theater technology is stealing fans from sports stadiums and movie theaters.

"One-trick ponies": PCs killed dedicated word processors. Now Apple and Google are talking about the "Post-PC" world. As Steve Jobs put it, "We're going to demote the PC and the Mac to be just another device," and "We're going to move the hub of your digital life to the cloud." Multifunction printers impacted sales of single-function fax machines and scanners. Siemens Healthcare has launched its Biograph mMR, which blends images healthcare professionals get sequentially from separate magnetic resonance (MR) and positron emission tomography (PET) machines. Smartphones have impacted single-function cameras, MP3 players, and GPS devices. The Evoz app is targeting the baby monitor. It lets you use one Apple device—an iPad, iPod Touch, or iPhone—as a monitor that stays in the baby's room and a second device as the receiver. Smartphones increasingly challenge medical devices. They could easily be the device that displays glucose levels measured by an adhesive patch with a tiny needle designed by DexCom, to stick to a diabetic patient's abdomen. Or display results from a wireless patch designed by Proteus Biomedical to work with pills containing microchips.

Those that ignore Moore's Law: VoIP calling is decimating overpriced landlines, long-distance, and mobile roaming from telcos. TextPlus and other services are making texting far cheaper and even free. In the meantime, the U.S. Congress has investigated wireless carriers and class-action lawsuits have been filed against them alleging that they conspired to fix texting prices. Auto dealers continue to expect thousands for integrated GPS and entertainment options when mobile/portable units are available for much cheaper.

Is Any Industry Safe?

Can any industry hide from technology-driven disruption? Some would argue the legendary investor Warren Buffett has done well by staying away from investing in tech companies (or those susceptible to technology-driven turmoil). In his words, "In business, I look for economic castles protected by unbreachable 'moats.' "[5]

Buffett prefers to stick with investments in companies like Coca-Cola because, as he says, it is "very easy for me to come to a conclusion as to what it will look like economically in five or 10 years, and it's not easy for me to come to a conclusion about Apple (or other tech company)."[6] And he says that in spite of his friendship and board counsel from Bill Gates, the founder of Microsoft.

Look at Coca-Cola, though, as it rolls out its Freestyle vending machines. "The dispensers each contain 30 cartridges of flavorings that mix up 100 different drink combinations. The cartridges are tagged with radio frequency ID chips, and each dispenser contains an RFID reader."[7]

Or Burlington Northern Santa Fe, another of Buffett's investments, which has used technology to lead the railroad industry, including "[a warning system that] uses satellites to detect speed-limit violations, improperly aligned switches, and missed signals. If a conductor doesn't respond appropriately, the system is supposed to stop the train automatically."[8]

Another example is Procter & Gamble, which is pioneering new forms of social media marketing. It is also extending its Tide franchise into the fragmented dry-cleaning business with stores that use email promotions, provide 24x7 secure locker access, "greener" cleaning options, and other technology innovations. The accomplishments of its CIO, Filippo Passerini, are described further in Chapter 10.

Next there is GEICO, one of Buffett's earliest and most lucrative investments. GEICO disrupted traditional car insurance by selling via mail rather than agents. Today, it is a technology marvel with elaborate call centers; many of its customers manage their GEICO interactions mostly by web or mobile device and its claim adjusters are equipped with Toshiba Toughbooks and digital cameras. In a likely Buffett-inspired knock on our technology-obsessed lives, GEICO is not, however, afraid to run commercials that show people doing dumb things with their smartphones!

No matter what your industry, it helps to be paranoid about the Netflix gunning for you. Of course, the rules of disintermediation keep getting nuanced. Apple has shown how a real estate intensive model in its retail system has been critical to its success. It is hugely dependent on labor-intensive Foxconn, a contract manufacturer with over a million employees in China, and its application ecosystem with applications and games from tens of thousands of entrepreneurs.

Offense, Not Defense

Instead of paranoia driving technology strategy, it is better to look at proactive opportunities to leverage the ubiquity of technology around us no matter what the industry.

It helps to think like Salvador Dali, who once said, "At the age of six I wanted to be a cook. At seven I wanted to be Napoleon. And my ambition has been growing steadily ever since." The new Dali museum in St. Petersburg, Florida, reflects that big thinking of the artist it honors. While its architecture and waterfront location are stunning, even more so is the design and technology to withstand the hurricanes that frequently threaten the state. The glassed areas are made up of 900 triangular-shaped panels, each unique. The areas holding art have 18-inch concrete walls with about 200 miles of reinforcing steel, and have been tested to withstand a Category 5 hurricane.[9]

Fountaindale Public Library in suburban Chicago, a model for the "library of the future," features flat-screen TVs, computer terminals, self-checkout stations, an automated book sorter, wi-fi, and other technologies.[10] Across the country in California there is an even more futuristic view: "Thus, the concept of a study center with computers, Wi-Fi, study tables, comfy chairs, and DVD and CD loans began to develop. The branch might not house stacks of books (it still could—we're still reviewing our options), but library patrons could 'order' books from the large Central Library (located about four miles away) and have them delivered to Marina Park the next day. This branch could be construed as a 'digital library,' but the Newport Beach Public Library system would have plenty of books and other printed materials readily available for borrowing."[11]

Other examples include:

- There are online classrooms. "Salman Khan, a former math geek and young hedge fund analyst, five years ago began making 10-minute videos to help his struggling nieces with their math homework. For convenience sake, he posted them on YouTube and found to his surprise that other people liked them, as well. Today, the Khan Academy website boasts 2,300 separate math tutorials, from simple addition to vector calculus, that have been viewed more than 50 million times by more than 2 million students and are in active use in more than a thousand classrooms across the country."[12]
- Take Kosaka, a unit of the metals-and-mining company Dowa Holdings. "One cell phone can yield up to 20 milligrams of gold; that may seem minuscule, but consider this: A ton of phones can provide 20 times more gold than a ton of gold ore. The company's recycling process is based on methods long used by Dowa to get metals from raw ore. Disused, dismantled electronics are heated to 1,300 degrees Celsius, at which point 19 different metals (so far) can be extracted. It's working on ways to harvest more. One target: neodymium, a rare-earth essential for magnets used in everything from microphones to wind turbines."[13]
- There's Mattel, with its "mental marathon"—the Mindflex Duel game. Using a brain machine interface from NeuroSky, players move a ball through an obstacle course, using only the power of their minds. Talk about future Jedi Masters and the ability to control events with a wave of the hand!
- There's the Lego-like videogame Minecraft, where players use blocks to create buildings and just about any other structure. Now

3D-printing software called Minecraft.Print is letting the game's fans bring their creations into the real world.[14]

- There's the Keeneland race course in Lexington, Kentucky. The 75+-year-old institution has been rolling out simulcasting of races, social apps, course technologies (sensors on horses, safer turf), and thoroughbred auction tools.
- There's Coldwell Banker, which is putting QR codes on real estate signage. The bar codes can be read by cameras on mobile phones to link to a web page on the phone's web browser. They give an interested party access to all kinds of information and often access to videos that walk through the property.
- There's Trademarkia, whose major technology innovation is a powerful search engine that mines the U.S. Patent and Trademark Office (USPTO) database more efficiently than the government's own search tool. "Trademarkia's business model is classic upselling: Individuals searching for trademarks via Trademarkia are urged to sign up for a trademark-application filing service with (co-founder) Abhyanker's law firm."[15]
- Then there are universities thinking ahead. "To capitalize on the growing cachet of the tech industry, colleges nationwide, including Stanford, the University of Washington, and the University of Southern California, have recently revamped their computer science curriculums to attract iPhone and Facebook-obsessed students, and to banish the perception of the computer scientist as a geek typing code in a basement. . . . Even universities not known for computer science or engineering, like Yale, are seizing the moment. The deans of the Ivy League engineering schools recently started meeting to hatch ways to market 'the Ivy engineer.' "[16]

The "Phoenix"—Technology-Driven Renewal

We have identified a number of industries where companies were demolished by tech-savvy competitors, but there are also plenty of companies that have reinvented themselves or commercialized technology that did not succeed in the first go-around.

Exhibit A is of course Apple.

In 1997, shortly after Steve Jobs returned to Apple, Dell's founder and chairman, Michael S. Dell, was asked at a technology conference what might be done to fix Apple, then deeply troubled financially.

"I'd shut it down and give the money back to the shareholders."[17]

A decade later Apple was valued more than Dell, and it has since gone on to even higher heights.

GoGo, the wi-fi service being offered by many U.S. airlines (like Delta, discussed further on, and Virgin America, discussed in Chapter 6) leverages Aircell's experience with airphones that we had in most seatbacks in the 1990s. Aircell paid the FCC about $30 million to acquire licenses for the air-to-ground frequencies. That allows it to differentiate from the satellite services of Row 44 that Southwest and others utilize. Planes transmit via underbelly blade antennae to 92 cell towers that can be accessed around the country and up to about 100 miles over international waters. In the long run, with numbers starting to build nicely (such as 1 GB of data per flight), the licenses may turn out to be one of the best investments Aircell has made since they are leveraging previous investment in the towers.

Delta Airlines has had a roller-coaster ride, along with the rest of the airline industry. In its heyday, the Delta Red Coat was a symbol of exemplary customer service authorized to do all kinds of favors for passengers. Delta has reintroduced the Red Coat, this time armed with Motorola handhelds to "help them more efficiently assist passengers, directing those who've missed a connection to their new flight, for example, securing boarding passes, or even providing food vouchers if there is a need."[18] Delta has also leveraged GoGo aggressively, says Cheryl Scheck, who leads eCommerce strategy for the airline. With more than 2,200 flights daily, Delta operates the largest fleet of wi-fi-enabled aircraft in the world. Increasingly, in addition to convenient flight times and air fares, the availability of wi-fi is a decision criterion for many passengers, and it provides Delta a competitive advantage. It was one of the first airlines to leverage Facebook for online check-ins. At selected airports, Delta has worked with franchisees to operate restaurants that are physically located in the gatehouse area and use iPads for customers to place orders. These same devices are free for customers to use to surf the Internet. This new "digital experience" is allowing Delta a fighting chance to regain its past glory.

Delta's other big competitor, UnitedContinental, has announced plans to move to a "paperless flight deck." Each pilot will be issued an iPad with a Jeppesen Mobile FliteDeck app. Jeppesen is a unit of Boeing and has traditionally delivered flight plans, navigation, and other data to pilots via paper. Both Delta and United are trying to stay current with a new standard in technology-enabled air travel being pioneered by startup carriers like Virgin America, which we describe in the case study in Chapter 6.

The GM OnStar telematics, with its navigation and emergency services, was a significant differentiator for its cars a decade ago. Then the world moved to Garmin and other handheld GPS units and to Google Maps on our smartphones. As we described in Chapter 1, GM is repositioning OnStar via a rearview mirror and on an annual fee basis so even Toyota and Ford customers can avail themselves of it.

Corning, as we describe in Chapter 15, has shown a remarkable ability to go back into its vaults and repurpose technology that was once ahead of its time. For example, the basic materials in its fiber optics had been available since the 1930s, but the market really took off in the 1990s when telecom companies started replacing their copper networks. Gorilla Glass, its very successful product for consumer electronic displays, got its inspiration from strengthened glass Corning had developed in the 1960s.

As the U.S. Postal Service looks to reinvent itself, it will find role models in a number of European peers. "Itella, the Finnish postal service, keeps a digital archive of its users' mail for seven years and helps them pay bills online securely. Swiss Post lets customers choose if they want their mail delivered at home in hard copy or scanned and sent to their preferred Internet-connected device. Customers can also tell Swiss Post if they would rather not receive items such as junk mail. Sweden's Posten has an app that lets customers turn digital photos on their mobile phones into postcards. It is unveiling a service that will allow cell-phone users to send letters without stamps."[19]

Then there is the history of Bayes Theorem, which influences so many technologies today. As Sharon Bertsch McGrayne writes in her fascinating book,[20] "respected statisticians rendered it professionally taboo for 150 years—at the same time that practitioners relied on it to solve crises involving great uncertainty and scanty information, even breaking

Germany's Enigma code during World War II, and how the advent of off-the-shelf computer technology in the 1980s proved to be a game-changer. Today, Bayes' rule is used everywhere from DNA decoding to Homeland Security."

Conclusion

Out of either paranoia or promise or "phoenix thinking," industry after industry is looking at next-generation technology-influenced products and services. It is almost impossible to find an industry with an "unbreachable moat" where technology's influence is not felt. Let's next look at the amazingly ambitious vision of leveraging technology even in a tiny place like Roosevelt Island.

Case Study: Roosevelt—Innovation Island

In the summer of 2011, Jonathan Kalkin submitted an entry in the One Prize Competition.[21] The competition invited creative ideas for linking New York City's five boroughs with transit hubs; expanding waterborne transportation; providing in-water recreation, educational events, and cultural activities; and promoting climate resilience. In April 2010, New York City Mayor Michael Bloomberg launched the New York City Waterfront Vision and Enhancement Strategy (WAVES), a citywide initiative that will create a new sustainable blueprint for the City's 578 miles of shoreline. Planning Commission Chair Amanda Burden had said, "Water is so important that we need to think of it as the sixth borough."

Kalkin is a resident of Roosevelt Island, which sits between the New York City boroughs of Manhattan and Queens in the East River. Kalkin recently served a three-year term as a Director of the Roosevelt Island Operating Corporation of the State of New York (RIOC).

His entry in the One Prize competition proposed a hybrid/solar fuel cell ferry, fuel cell buses, bike sharing stations, and interesting ideas for smart parking.

What business does a tiny island only 147 acres in size and with a population of just over 13,000 have thinking that big?

For starters, it has Kalkin, who is a self-described geek and an optimist in technology, being able to solve almost any problem. Kalkin says he spent most of his childhood designing and building projects the size of his room with Construx (the Fisher-Price plastic building toys) and using computers and the Internet at a young age to track and pick investments so he could save up enough to buy a robot. The company that made the robot went out of business by the time he had enough funds, but he used this knowledge in design and technology to help him as a principal in the financial industry and later as a director at RIOC.

A Track Record of Technology Innovation

Kalkin started from a good base. The island has the only aerial commuter tram in the country. The tram takes residents across the river to

Manhattan in just four minutes. Kalkin oversaw the state-of-the-art tram redesign in 2010 as the Chair of the Operations Committee. Verdant Power has been using the river during ebb and flood to generate tidal power. It has an initial demonstration array of six turbines to grow to a planned full field of units that could generate up to 10 MW of power.[22] The island has an automated vacuum collection system: Residents drop trash down a chute, anywhere on the island, and the garbage, depending on the density, shoots through pneumatic pipes at 30–60 miles per hour.[23] That means no messy streets, and importantly, no garbage trucks clogging the narrow streets of the island. Each of these technologies has been operational for years. The garbage system started in 1975, the aerial tram in 1976, and the tidal power generation in 2002.

And the RIOC board has been busy in other ways since Kalkin joined in 2008. Free fiber wi-fi is coming first to the island's parks, with a second phase wrapping most of the island courtesy of Verizon. Verizon is also gearing up its FiOS broadband service for local residents and businesses. The Red Buses that meet the tram described earlier and take residents across the island are tracked real-time by NextBus. It uses GPS and predictive analytics to inform residents via their smartphones or by text message of the status of the bus. They can decide whether to wait for the bus or walk to the tram. In December 2011, Mayor Bloomberg announced a "transformative" event for the Island. Cornell University, partnering with the Technion-Israel Institute of Technology, plans to build a $2 billion Applied Science school there.[24]

Smart Parking

More significantly, the RIOC has tackled the parking congestion issue by bringing in a vendor called Streetline. Streetline brings together all aspects of parking management technology in one integrated system—from street-level sensors to mobile apps, analytics, and system management software. This helps cities more effectively manage their parking resources and helps motorists find parking more quickly and efficiently.

Zia Yusuf, president and CEO of Streetline, says, "Roosevelt Island is setting the stage for a new world of efficiency and improved services for its citizens. Smart parking solutions can significantly reduce traffic and congestion in cities, improving air quality and in turn quality of life."

Experts estimate that 30 percent of urban traffic is caused by motorists looking for parking. Additionally, vehicle emissions and drivers looking for parking are so closely linked that a yearlong study found that drivers in a 15-block district in Los Angeles drove in excess of 950,000 miles searching for a space and produced 730 tons of carbon dioxide.

Streetline's patented ultra-low-power sensors and Meter Monitors communicate via a wireless mesh network to deliver valuable real-time information. It notes how long a car is parked, when a car enters and leaves a parking space, and whether payment has been received for metered parking areas.

Kalkin elaborates:

> We looked into smart parking technology because, like most urban areas, we have limited parking, but want people to have the ability to park on the street. They may want to visit a friend, to get something to eat, or drop things off at their apartment. We also have a long-term parking facility on the island.
>
> There is a school of thought that dynamic pricing works best for parking. Instead I believe that proper enforcement, coupled with lower prices for short-term parking on the streets and longer-term parking in our indoor garage, would actually minimize the parking abuse on the street and allow more spaces to be readily available for their intended purpose.
>
> Readily available parking is good for the local economy and minimizes traffic congestion. It also helps our Red Bus stay on schedule, because when people have parking spaces on the street, they don't double park.
>
> This also gives us a much more efficient public safety force. Sensors can do a much better job than humans tracking events like people overfeeding meters or not paying at all using fake parking placards. The police are also using the (Streetline) enforcement features with their smartphones. As mobile phone enforcement evolves, a smartphone can become a one-stop resource to document crimes or violations through pictures or video, writing of tickets, and scanning of parking placards with QR codes to check for validity. Our police can also track and answer 311 (non-emergency version of 911) type violations with our

> SeeClickFix citizen reporting system. All that allows our public safety force to focus on more important tasks, like keeping the community safe, and automates this enforcement process.

On the consumer end, mobile parking applications can provide consumers with parking information before they arrive. On Roosevelt Island, for example, motorists can use Parker by Streetline to quickly find and pay for available parking spaces via their smartphone.

> With location-based technology evolving, you should start to get special offers on your phone from retailers close to where you park. With mobile payments we are reducing the need to step out and get a parking ticket. The easier you make it to pay for parking the more likely people will do it. Also with signage on the street and real time parking information, you allow the consumer to know when and where they can park. Today the parking uncertainty causes people to hesitate visiting the island. As electric cars grow in popularity, it will be very useful to know which spaces are available with charging. And they tell us about patterns we need to assess as the Island population changes. There is a lot more to smart parking than being able to find a space. . . . it should be a boost for the local economy.

Early data from the smart parking plan is still being analyzed, but RIOC has noticed a positive difference in parking meter payment, parking law compliance, congestion, and availability compared to areas where the sensors haven't been installed.

Which brings us to the competition and Kalkin's entry.

The One Prize Competition Entry

Kalkin describes his competition entry:

> The basic concept is to create a transit hub where ferries, buses, cars, and bicycles can converge and use the tidal energy we have already been leveraging and adding to electric power from solar and wind sources from the roof. This will eliminate the common argument that electric transportation is not really green because the power comes from fossil fuels. It also will consolidate several forms of transportation in one area, which saves energy and is more convenient to citizens.

> Tidal energy is renewable, clean, tested, and readily available in New York City. It is more reliable and predictable than solar or wind.
>
> Our concept introduces hybrid solar/fuel cell ferries, which are increasingly becoming popular around the world. One of the main concerns of ferry systems is that they have low or no profit margins and are not subsidized like most public transportation. However, the price of diesel fuel is high and therefore it is difficult or impossible for current types of ferries to compete with other forms of public transportation even if they are more convenient for some and take pressure off subways, buses, and car traffic. Also, communities are reluctant to have docks near their buildings because of noise and air pollution when ferries are idle.

New York City is investing $9.3 million over three years to subsidize a new conventional fuel ferry service, which started in June 2011—but it does not service Roosevelt Island.[25]

Kalkin continues:

> Hybrid/solar ferries use 30 to 40 percent less fuel and can be plugged in overnight with no emissions while idle at the station. This charging station would be powered by tidal energy. Hydrogen can be separated from the river water to power the fuel cell through a process called electrolysis. We would also leverage the solar and wind energy generated on the roof. The green savings from this could be used in place of the usual transportation subsidy and make the ferry service a competitive/potentially profitable and worthwhile addition to a community. The typical initial government subsidy can instead go into retrofitting present ferries so the cost and initial investment is kept to a minimum. The EPA currently provides an almost complete subsidy for certified retrofits of marine vehicles through their emerging technology program.
>
> The next generation of Red Buses would use hybrid plug-in fuel cell buses that are entering the market. These buses, like the ferries, would charge onsite using tidal energy and the hydrogen extracted from the water through electrolysis and the solar and wind from the roof of the transportation hub.
>
> The floors of the parking garage can be made "charge station ready" and the power from the stations can come from tidal energy. If a person moves (and in New York City, people move quite often), they can take their portable Level 2 charging station (like the Leviton

Evr-Green series[26]) with them. As level 3 charging stations become the standard, the transformer on each floor simply has to be modified, because most parking garage lighting has the same or similar voltage of level 3 stations. This is a scalable and future-proof solution to a city parking garage that can be upgraded by floor rather than by each parking spot and charging station with little change to the electrical infrastructure.

The parking garage in the hub would also be a place to service/convert and sell electric cars. As an example, used hybrid plug-vehicles (say from Zipcar) can be sold after two years of shared use onsite. Used gasoline vehicles with engine issues can be donated for green transit credits and converted to plug-in hybrids like the successful Ford Escape plug-in conversion and/or converted to run on biodiesel in the "conversion shop." The conversion shop in the garage and the tools would also be powered by tidal and other renewable energy. Or the converted vehicles could be used for car shares to gain a second life. When renting the vehicle for a car share, the person would be given the electronic option to be put on a waiting list to purchase the car when it reaches the end of its two-year rental cycle. We can get away from wasteful test drives and vast car dealer showrooms.

Another concept in our proposal is a bike-sharing station. You could use them as stationary bikes, and when pedaled the energy would be used to create power for the hub. In addition to calorie counters to show how long you have exercised, we would reward you with credits on the Project Verde card described below based on reduced carbon emissions. The station would be located on the roof of the parking lot, so users can enjoy views of the harbor. Of course, you could also remove the bike and use it to ride around the island.

Each user would purchase a Project Verde RFID (wireless) card and wave it to pay for any transportation in the hub. Use of each kind of transportation in the hub leads to Verde credits, which are higher the "greener" the activity you are doing. So using the hybrid solar ferry or using the bike sharing in stationary mode creating power would net you more credits than using the car-share service. In addition, using transportation in off-peak periods could gain more points and reduce traffic congestion at the hub. Points could be used to get free or reduced fares on any form of transportation and for use with participating local business. This would create an incentive to stimulate the local economy. Of course, we don't necessarily need a separate card—it could be integrated with credit/debit cards or growing mobile payment options that allow people to pay for transportation with their phone or wirelessly wave it over a sensor at a transportation gate like an electronic ticket.

The exciting thing is that this concept can be replicated beyond Roosevelt Island to every other ferry stop along the East River. Think of how green that would make New York City.

Turn a Vision into Reality

Of course, at this point, it is just a proposal in a competitive event. But Kalkin is excited about implementation opportunities: "There are a number of funding possibilities. When looking into a project, I always try to ask the question, what will this cost us over time if we don't innovate? First and foremost, most forms of transportation are subsidized. Green transportation has an initial higher cost, but over time the fuel savings can pay for the initial investment. There are grants to offset this initial cost, and it can actually cost the government less money over time because there is no need or less need for an ongoing transportation subsidy. In addition to reducing pollution and adding high-tech jobs to the economy, technology helps government become more efficient and responsive to citizens in the community. It reduces parking congestion and improves public transportation reliability which can help revitalize the local economy around it and increase ridership which, in turn, increases government revenue. This efficiency allows government to perform services better with less money and hours spent on manual tasks that are automated. The end result is that citizens and the local economy get the environment needed for business growth and a more responsive and cost efficient government."

Tiny island, mighty goals. After all, President Franklin D. Roosevelt, who the island is named after, said in a commencement address, "The country needs and, unless I mistake its temper, the country demands, bold, persistent experimentation. It is common sense to take a method and try it: If it fails, admit it frankly and try another. But above all, try something."[27]

Chapter 4

Australia to Zanzibar: No Country for Old Products

"Designed by Apple in California."

Those five words are embossed on countless devices around the world. Those are words for consumers to lust after, but they mask a global journey beyond their initial design in Cupertino, California:

> Manufacturing iPhones involves nine companies, which are located in the PRC (People's Republic of China), the Republic of Korea, Japan, Germany, and the U.S. The major producers and suppliers of iPhone parts and components include Toshiba, Samsung, Infineon, Broadcom, Numunyx, Murata, Dialog Semiconductor, Cirrus Logic, etc. All iPhone components produced by these companies are shipped to Foxconn, a company from Taipei, China located in Shenzhen, PRC, for assembly into final products and then exported to the U.S. and the rest of the world.[1]

That was iPhone version 1.0. Later Apple phones and devices have used different suppliers and different countries.

On the customer front, when Apple introduced the 3G version of the iPhone in 2008, it did so in 22 countries, with more than 50 other countries following over the next few months. The excitement was universal:

> In Hong Kong, designer Ho Kak-yin, 31, wearing a T-shirt that said, "Jealous?" was first in line in a queue of about 100 inside a Hong Kong shopping mall. "I'm very excited. It's very amazing," Ho said, after lining up two hours ahead of the kickoff. Hundreds queued outside stores in New Zealand's main cities got their iPhones earlier at midnight Thursday.[2]

The rollout required coordination with telcos like VimpelCom in Russia, Etisalat in Dubai, and Telia in Sweden, which many of us have never heard of.

Fast-forward to the iPad Apple introduced in 2010.

Mike Laven's bio modestly states he "has had a long career in financial technology in both Silicon Valley and London." Starting with his MA in International Affairs from The School for International Training at Harvard and a Peace Corps assignment in India, and numerous technology jobs in between, he is equally at home in Paris, Delhi, and Tel Aviv. He truly is a "global citizen."

He describes a scene at a spa in June 2011 in southern Germany:

> Well-off clientele and always totally global: Egyptians, Russians, Israelis, Lebanese, the Stans, smatterings of everywhere else, many Europeans and rarely an American. The universal adoption of the iPad this year is incredible—everyone has one, ages 25–70. People are doing mail, checking local newspapers, getting news from home, but this is not a surfing/gaming crowd. Reminds me of last year's universal adoption of the Stieg Larsson novels, when everyone was reading the same book in different languages. Interestingly, I don't see many of the other tablets which I see in the U.S. While everyone else is talking Apple has locked up global opinion makers.

Apple's global prowess is impressive, but it is a mature company over three decades old.

How about Groupon getting to almost $650 million in quarterly revenue as a two-year-old startup?[3] It has been called the fastest-growing company of all time. It bought stakes in similar coupon/deal sites around the world, including CityDeal in Germany, Darberry in Russia, ClanDescuento in Chile, and Qpod in Japan. It offers daily deals in over 500 cities around the world.

Or Netflix. It was primarily a U.S. player until it expanded into Canada in 2010 with modest results as its streaming service was hindered by low data caps by the country's broadband ISPs. That led people to believe it would cautiously expand globally. To the surprise of many, it announced plans to enter Latin America in a big way!

> While a populous and fast-growing part of the world, these aren't generally countries considered broadband leaders. For example, according to OECD data of per 100 inhabitant broadband penetration as of December, 2010, just two countries in the region—Mexico and Chile—fell into the top 34 in the world (#32 and #33 respectively), each with a 10.4 percent penetration rate per 100 inhabitants.[4]

With this blitzkrieg approach to global expansion, Groupon and Netflix are clearly betting that if they don't put down stakes, others will do so.

Some would say Apple, Groupon, and Netflix reflect the realities of Tom Friedman's description of a new Flat World.[5]

"Globaloney"

Actually, the world is not that flat. Pankaj Ghemawat, in his book *World 3.0,* calls that view "globaloney."

Why?

"Because economic data simply don't support the view that we live in a flat, connected world, even if we are technologically connected with everyone, everywhere, all of the time. Data show that most types of economic activity that could be carried out across national borders are actually still concentrated domestically."[6]

Look at the world from the eyes of someone "down under." Ben Kepes is an analyst with Diversity, Ltd. in New Zealand:

I live a life full of dichotomies. I live in rural New Zealand in the midst of an idyllic valley and surrounded by vineyards and olive groves. My working life however sees me visit the U.S. several times each year, and spend most of my working day interacting with colleagues overseas primarily in the U.S. but in other countries as well. This dual life gives me an interesting perspective on technology and its adoption in different countries.

Case in point is broadband. New Zealand has a single fiber optic cable connecting it with the outside world. As would be expected, this single cable (or more correctly, the single provider that sells access to the cable) means that data caps, poor access speeds, and throttled capacity are the norm. While moves are currently underway to lay an alternative cable between New Zealand and the outside world, it's fair to say that the realities of limited connectivity have had an impact on the uptake of technology. Given that the vast majority of data our citizens access come from outside our own borders, it never surprises me that smartphones are still a relatively unique phenomenon in New Zealand. Compare that to Silicon Valley, where it's rare to see someone without a data-enabled mobile device.

It is perhaps for this reason that New Zealand is one country where eBay hasn't realized market domination of the online auction space. Rather a local site, TradeMe, which not only has a local flavor to it but, more importantly, serves up content from local sources, is by far the biggest player in this space. In fact TradeMe is widely credited as being the service that really brought Internet usage to the masses. This amazing uptake saw TradeMe sell to Australian media group Fairfax for some $700 million several years ago. People have commented that TradeMe was helped to its position of strength by relatively slow uptake of technology caused by connectivity constraints.

In other areas, however, where connectivity isn't such an issue, we're real leaders in technological innovation. Witness EFTPOS, our own debit card system for electronic transaction processing. EFTPOS (short for Electronic Funds Transfer at Point of Sale) was introduced here in 1985 as a pilot scheme in gas stations. Since that time EFTPOS has become very much the de-facto way of paying for goods. The majority of all retail transactions are processed using EFTPOS, a figure that puts New Zealand close to the top of global debit card usage. In fact, a random observation at convenience stores, gas stations, and other retail outlets shows that New Zealand has almost attained that long-predicted status of a cashless society.

In such an uneven world it helps to have a "human-behavior researcher" like Jan Chipchase, who carries his Nikon around the world to send pictures back to Nokia designers in Finland and elsewhere:

> (Chipchase) could be bowling in Tupelo, Mississippi, or he could be rummaging through a woman's purse in Shanghai. He might be busy examining the advertisements for prostitutes stuck up in a São Paulo phone booth, or maybe getting his ear hairs razored off at a barber shop in Vietnam.[7]

Local Nuances

Learning global nuances unveils nuggets like the following ones:

- A software company executive visiting Poland expects to review development opportunities there. Instead, he is presented with a market opportunity. "Your customers in the United States and Western Europe are hesitating to adopt your cloud computing offerings because they have to integrate with plenty of old legacy systems. We have little legacy—so few barriers to adoption. Come here first."
- Then there is what is being called "trickle-up" innovation, or reversing the traditional global rollout model. GE Healthcare introduced its MAC 800 electrocardiograph machine in the United States for a mere $2,500, an 80 percent markdown from products with similar capabilities. But what really distinguishes the device is its lineage. The machine is basically the same field model that GE Healthcare developed for doctors in India and China in 2008.[8]
- A GE competitor in medical devices, Medtronics has a joint venture with Chinese firm Weigao for several reasons:

  > First, local demand: As China's second-tier cities boom, their clinics are crying out for cheap gizmos. Second, the Chinese government is pushing 'indigenous innovation' by favoring local firms in tendering, procurement, and so on. An engineer gleefully points out that as a local entity the Medtronic joint venture can

buy essential rare-earth metals cheaply, despite Chinese restrictions on their sale.[9]

- There are unique regional nuances:

  —The Japanese lead the world in their use of vending machines and service robots:

  > Japanese people have been exposed to robots as heroes that bring justice and peace. On the other hand, in the United States and Europe, robots are often depicted as enemies. Vending machines are one kind of robot, so Japanese people accepted this robot as easily and subconsciously as they accepted other robots in cartoons.[10]

- Few mobile users in India use voicemail and instead have evolved their own version of signals around "missed calls":

  > Missed calls are when you call someone you know, you let it ring only once or twice, and then cut the call. This is generally between people who talk regularly and a missed call conveys a preset message. For example, a wife might give her husband a missed call at the end of the workday to convey that she is heading home and that it's time to meet up. Or someone might give a missed call to a friend they commute with indicating that it's time to pick him up. Or a student might give a missed call to her parents—so that the parents can call back (that way the parents incur the cell phone charges).[11]

- The "Arab Spring," the revolts that spread across Tunisia, Egypt, and other parts of the Middle East, opened eyes to the impact of social networks even in authoritarian countries and the marketing opportunity they offered. There are

  > 7.4 million Facebook users in the Gulf according to Inside Facebook as of May and 5.5 million Twitter users in the Middle East according to ArabCrunch as of March (of 2011).[12]

- Think only the Swiss and the Japanese can make precision watches? London-based Hoptroff is a "silicon foundry that designs and patents electronic watch movements for 'future classic' timepieces."[13] Hoptroff continues the tradition of other watch industry innovations from that city, including the 1664 invention of the balance spring and the 1753 innovation related to temperature compensation. An example from their "Carte des Ebauches":

> Bluetooth Low Energy is a new wireless technology being introduced during 2011 and 2012. It allows the wristwatch to make the small communications hop to smartphones while still only requiring coin-cell power. The phone connection allows the watch to complement the phone, rather than compete with it for the customer's attention, and allows access to live data from RSS data feeds using Hoptroff software on the phone.

- There's Spain, the land of Don Quixote and his tilting against windmills, now a global leader in wind energy. Iberdrola Renovables, the world's biggest producer of wind power by installed capacity, is based in Valencia but operates in over 20 countries.

All this has significant impact on global product launches and marketing considerations. It is very easy to ignore promising markets and focus on traditionally "rich" countries. On the other hand, it is easy to assume consumers in each market behave the same. Public policy also differs considerably around the globe and regionally within countries as we see in the next section.

Impact on Public Policy

In the United States six years ago, just two markets—the western and southeastern regions—accounted for more than 55 percent of the nation's renewable energy generation capacity. "Their share of capacity is now down to about 40 percent; other regions have grown at a faster clip."[14]

Two-thirds of Iceland's population (approximately 320,000) is on Facebook,[15] so the council that is revising its constitution broadcasts its weekly meetings live not only on the council's website, but on the social network as well.

Singapore has turned itself into a laboratory, teaming with MIT to form the Singapore-MIT Alliance for Research and Technology (SMART) center to examine the "future of urban mobility" as well as other growth issues. Its purpose? "To study how cities work and how they can work better."[16]

Joe Garde of IrishDebate.com points out the Irish are the "glue" in many enterprises around the world. This has allowed the Irish Diaspora outside Ireland, estimated at 20 times the population in the island, to thrive. It challenges the Irish government, though. Should it prepare its young talent to "export," as it did for decades, or should it find a way to attract multinationals to invest there as they seek tech-savvy talent and replicate its more recent Celtic Tiger success?[17]

South Korea is investing 2.2 trillion Won ($2 billion) under its Smart Education program to digitize all elementary and secondary school textbooks. This way they can be read on a variety of devices, including computers, interactive whiteboards, iPad-like tablets, and smartphones. Classes will also be video-streamed online so children who can't come to school due to poor health or weather don't miss out. Children with disabilities may also benefit: e-books could be controlled by eye-tracking or gesture recognition, for example.[18]

Quincy, in a rural part of Washington State, has quietly become a data center hub for companies like Dell, Intuit, and many others. "Quincy officials say the data center building boom is helping them buck the trend. Real estate prices are rising, and tax revenue surged to $2.24 million last year, from $700,000 in 2005, according to City Administrator Tim Snead. Quincy recently spent $1 million on a new library and $75,000 for a new museum parking lot. It repaved 60 percent of the streets, bought a hook-and-ladder truck for the fire department, and erected an edifice for such services as the management of parks and water."[19]

Of course, while we compete on earth, there are plenty of opportunities to collaborate, especially in space. The International Space Station has had over 200 people from 11 countries over the past decade, and it has been assembled from components named Destiny, Kibo, and Zarya from 15 countries.

And Yes, Let's Not Forget about Zanzibar

That's Zanzibar as in the bar that Billy Joel made famous with his hit in 1979 (it included a memorable solo by the famous jazz trumpeter

Freddie Hubbard). The hospitality and restaurant industry has a continuing problem with inventory accuracy and losses. It is an established industry fact that approximately 25 percent of all alcoholic beverages disappear, and not via evaporation. Automation to help control inventory has been of limited success, mostly due to intrusive methods for setup and control or because the overtaxed bar staff were required to do additional tasks to manually facilitate inventory control. Beverage Metrics Inc., a California subsidiary of Austrian company Identec Group, has built a solution utilizing second-generation wireless technology to provide real-time control of inventory in general and beverages in particular. Additionally, its solution can reconcile actual liquor, beer, and wine pours with point-of-sale (POS) system entries without any additional tasks being performed by the bartending or waitstaff of a restaurant, bar, or banquet. This solution also provides analysis on bar staff performance around over- and underpours, substitutions, compliance with mixed drink recipes, and compliance to a POS revenue transaction for every pour dispensed. The system uses sensor-enabled RFID tags and readers. They are capable of sensing the pouring motion of the liquor or wine bottle and communicating to the middleware infrastructure. That translates the sensor data into an estimate of the volume that was poured. This "pour engine" is 90 to 95 percent accurate, considered the best in the industry. An in-line turbine system is used to accomplish the same task for draft beer.

Conclusion

In some ways we live in a flat world where Apple, Google, and Facebook have become our common language. On the other hand, there are so many unique nuances that it would be naïve to assume global branding and universal product versions can be successful for all products. The different rates of technological evolution also offer significant policy opportunities as states, provinces, and nations compete for jobs and investments. Let's next look at Estonia's remarkable "Tiger Leap" and rapid evolution away from decades of communist stagnation.

Case Study: Estonia's "Tiigrihüpe"—Tiger Leap

"George Washington."
"George Washington."
"George Washington."

Rein Krevald is using a heavy accent as he mimics his late grandmother's answers to every question on her U.S. citizenship test. The Immigration Officer had every reason to fail her. Instead, he chuckled at the grace of the lady and said, "Welcome to the United States."

Krevald, born in the United States, continues reminiscing about his family with roots in the tiny Baltic country of Estonia.

> It occurs to me that my parent's success when they arrived in the United States in 1947 and Estonia's success after the fall of the Soviet Union in 1991 had a very significant common denominator: Starting with a blank slate!
>
> When my parents arrived in the United States, they had absolutely nothing material to their names but they both had an education. My mom was a pharmacist and my father had been a lawyer in Estonia. It was almost impossible for them to aspire to their previous careers in this country. Being open minded towards any opportunity that came along, my mother launched a successful career as a cosmetic chemist (she ended up with numerous patents to her name) and my dad ended up working for CitiBank in New York.
>
> When Estonia regained its independence, it also was starting from scratch.... Estonians had suffered during the Communist regime (even though the Soviets had secretly assembled their first computer and designed their first space mission in an Estonian research park), but to the best of their ability had kept up with the rest of the world especially through watching Finnish television.

The capitals of Finland and Estonia are just 50 miles away as the crow flies. Before the Soviet takeover of Estonia, the countries had roughly similar standards of living. Over the next 50 years, the Finnish per capita income was estimated at seven times as much as that of Estonia.

> Same with the country after the Soviets left—decisions had to be made quickly but starting from square one made it easier.

Jaan Tallinn, one of its founders of Estonia-born Skype, agrees with Krevald. Skype is the company that revolutionized VoIP calling and that Microsoft bought in 2011 for $8.5 billion. He says: "Because we started anew, we got new laws, new leaders, and new technology . . . . the big winners were the start-ups."[20]

Continues Krevald:

> Estonians were not mired with the existing infrastructure. Do we try to fix the crappy Soviet landline phone system or do we embrace cell phones? Easy answer. In no time almost every Estonian was using a cell phone. It helped that Nokia, a Finnish company, was close by.
>
> Western economists give a lot of lip service to a flat tax system. Estonia just went ahead and successfully implemented it.

As Erich Follath wrote in 2007: "Few countries are as crazy about the Internet as Estonia, and no capital city can keep up with Tallinn on that count. All schools are connected to the Internet; more than 90 percent of all bank transactions are conducted online; and there are more mobile phones than residents."[21] "You should see their electronic voting. Not like our hanging chads," jokes Krevald, who lives in Florida.

One of the architects of this digital society is current president (then Estonian ambassador to the United States) Toomas Hendrik Ilves, who in 1996 proposed a "Tiger Leap" for the country. The tiger moniker was in recognition of the "Asian tiger" economies that had blossomed using technology.

Krevald grew up with Ilves and was in the same Estonian Boy Scout troop in northern New Jersey. "You could tell he was into technology even back then. He and his father would be working on radios and electronics when the rest of us were running around in the woods playing with sticks."

The initial Tiger Leap focus was on education. Almost all the schools were supplied with computers in 1997, and a quarter of the teachers was trained on them. By the end of 1999 there was an average of one computer for every 28 students in Estonia, bringing its target of "one computer for every 20 pupils" within reach. More than half of the total number of teachers has graduated from the Tiger Leap's beginner course.[22]

Says President Ilves, "Every student's access to a computer and Internet is as natural today as having a lamp in the ceiling of the classroom."[23]

Today, of course, it is way beyond education. "For citizens of Estonia, e-services have become routine: e-elections, e-taxes, e-police, e-healthcare, e-banking, and e-school. The 'e' prefix for services has almost become trite in the sense that it has become the norm."[24]

Tiina Krevald (Rein's second cousin) has been in technology since 1994 even before the Tiger Leap initiative. She now works for Microsoft in Estonia, and describes her family's digital lifestyle:

> —We have free wi-fi everywhere. In practically all public establishments, from hotels to gas stations, there is a public wi-fi that is free of charge or for a small fee. I travel around the world, and nowhere is web access so easy—and usually it costs so much.
>
> —I can do all my banking online 24x7, even sign contracts digitally with my resident ID card (the card has a chip that not only holds information about the card's owner, but also two certificates, one of which is used to authenticate identity and the second to render a digital signature).
>
> —My tax declaration took five minutes this year. Everything popped up on-screen prefilled (my income, my donations, my tax exemptions for my young kids). I only had to check the accuracy of data and press the confirmation button. Shows the power of a flat tax, and the level of automation of the tax system.
>
> —Took me another five minutes to cast my vote. In the 2011 parliamentary elections 140,000 voters (25 percent of the total) used this convenience.
>
> —I pay for parking using my mobile phone. The charge shows on my cell phone invoice. It's available all over Estonia in public and private parking zones.
>
> —Our medical and real estate records are digitized and very easy to access in our state registries.
>
> —Via e-School I can check on my kids' grades, their absence from classes, the content of lessons, and homework.
>
> —Our prescriptions are digitized from the doctor to the pharmacy—so no lost paper scripts, or handwriting transcription challenges.

In 2007, the risks of that much digitization of the economy showed up. *The Economist* wrote, "for the past two weeks Estonia's state websites (and some private ones) have been hit by 'denial of service' attacks, in which a target site is bombarded with so many bogus requests for information that it crashes.

The Internet warfare broke out on April 27th, amid a furious row between Estonia and Russia over the removal of a Soviet war monument from the center of the capital, Tallinn, to a military cemetery."[25] *Wired* magazine called it "Web War One":

> All major commercial banks, telcos, media outlets, and name servers—the phone books of the Internet—felt the impact, and this affected the majority of the Estonian population. This was the first time that a botnet threatened the national security of an entire nation.[26]

NATO responded by establishing a Cooperative Cyber Defence Centre of Excellence (NATO CCD COE) in 2008, and located it in Estonia.

Having blazed a digital trail, Estonia is now poised to also lead the world on cybersecurity even as the world tries to emulate what Estonia has offered its citizens for over a decade now. It has set up a cyber National Guard. President Ilves notes, "It's a government-funded, white-hatted hacker organization . . . . People spend their weekends or evenings and they do something defense-related."[27]

President Ilves looks ahead and says, "We must give the Leap new meaning so we may cope with ever-changing functions." He honors "Tiger Achievers" on a regular basis and keeps encouraging entrepreneurs who are following Skype's lead and companies like Microsoft to expand there with talent like Tiina. Along with its Baltic neighbors, Latvia and Lithuania, Estonia is developing a nice reputation for mobile, social, gaming, and other applications.

In the meantime, Estonian citizens and Diaspora around the world are justifiably proud of the digital prowess. Some like to call it "e-Stonia."

Indeed, Krevald tells an Estonian joke that goes:

> At an archeological dig, Russian scientists found traces of copper wire and announced their ancestors had the world's first telephone network. Germans later announced they had found traces of fiber-optic cable

during a dig and concluded that their ancestors already had an advanced digital network. Soon after, Estonian newspapers reported a dig in Narva had found absolutely nothing. We, therefore, have concluded that our ancestors were the first users of wireless technology.

Chapter 5

Convergence, Crossover, and Beyond

Traditional technology buyers are learning to become technology vendors, as we saw in Chapter 1. Many vendors are replacing buyers as the new best practice leaders, as we saw in Chapter 2. Welcome to the world of the technology "buyor"—the switch-hitter, who is good at both sides of the baseball plate, like the legendary Mickey Mantle.

Ambidexterity is a skill valued in many other sports. There are the soccer players who can fire rockets with either leg and cricketers who have mastered the "reverse sweep." They look as elegant as Bjorn Borg did, when he dazzled the tennis world with his two-handed backhand on his way to multiple Wimbledon titles. They are the drummers who can use all four limbs to magical effect in a band.

Not just in sports—Michelangelo is supposed to have painted the Sistine Chapel with both hands (and it still took him years).

Ambidexterity in technology can be just as dazzling.

The "Buyor" Phenomenon

We saw UPS as "part transportation, part technology" in Chapter 1. There are others like 3M, GE, and The Washington Post Company described below.

Go to the 3M website and you see a "periodic table" with 46 elements.[1] Click on Bi, and it describes 3M in Biotechnology, Mf describes what it does with Mechanical Fasteners, and We with Accelerated Weathering. The company many of us know best for Post-it notes and Scotch tape has more than 55,000 products and has an uncanny ability to combine highly innovative technologies in new and unexpected ways. We cover the periodic table more in Chapter 18.

Then there is GE. "In sector after sector, we find that technology suppliers sometimes lack deep domain knowledge when it comes to vertical technology solutions. That has opened the door for GE Healthcare, GE Transportation, and other units to become technology leaders in their markets. We are a multibillion-dollar software and technology company in our own right."[2]

Donald E. Graham, Chairman and CEO of The Washington Post Company, which dates back to 1877, says it has become easier for companies to become technologically ambidextrous in the past decade. While traditional media has been decimated by technology, the Post has become a diversified technology company. Its Kaplan unit is one of the largest online education companies. Its Cable One unit is one of the largest cable companies in the United States and services a number of rural communities. It has digital properties like Slate. It has invested in startups like Avenue100, which provides marketing analytics to institutions. Trove (previously iCurrent) is a social news site and aggregator, and SocialCode helps companies with Facebook advertising and user engagement campaigns.

When the *Washington Post* launched its iPad app in 2011, it created a humorous video of a befuddled Bob Woodward being pulled away from his trusty typewriter and asking his editor, Ben Bradlee, about the allure of the iPad. It was a throwback to the huge influence the newspaper had in the 1970s in bringing down the Nixon presidency with its investigative coverage of Watergate.

Graham also sits on the board of Facebook and is arguably more tech-savvy than the average CEO or chairman. Modestly, though, he says it has become a lot easier for companies to hire technology talent like Vijay Ravindran, his Chief Digital Officer. Ravindran was involved with Amazon in its formative years and with other technology startups.

Graham's view: "Increasingly we can offer talent like Vijay equally challenging problems to solve as do technology vendors."

Vice versa, technology vendors have been hiring from the corporate world to streamline their own operations. "The conventional wisdom on high-tech vendor CIOs hasn't always been of the highest order. There are but a few 'rock-star' CIOs leading high-tech vendors' IT efforts. Most notable are HP's Randy Mott and Microsoft's Tony Scott. However, when the right CIO falls into the right situation, the combination can be powerful for both vendor and customer: a peer CIO who knows the lay of the land (IT governance, project management, and political challenges that IT leaders face) as well as the guts of the vendor's software (what it can and cannot do). It's the proverbial 'eat your own dog food' situation, espoused by many technology leaders."[1]

It Helps to Start Early in Life

If you talk to sports coaches, they will tell you "the earlier (in life) you start switch-hitting, the better." It is a lot harder later in life to get the upper-lower body coordination needed to be comfortable hitting a baseball or kicking a soccer ball from either side.

In May 2011, The Thiel Foundation announced fellowships to 24 recipients "younger than 20 years old at the end of 2010—each will receive $100,000 and mentoring under the condition that they stay out of school for two years to build their businesses. . . . The grant winners are pursuing a range of businesses, from alternative energy to education and e-commerce."[2] That's Thiel as in Peter Thiel, co-founder of PayPal and successful investor in a string of successes such as Facebook, LinkedIn, and others.

Thiel clearly subscribes to what the coaches say—"start them young." The Thiel fellowship's requirement that the winners stay out of school for two years is, of course, controversial. Indeed, the mother of one of the winners, who will leave Harvard, was quoted as saying, "This is a different paradigm from what I grew up with."[3]

Of course, talk to corporate IT folks and they will smirk at a budget of just $100,000. Chump change in a world of trillions in IT budgets.

Take a look at Spanning, based in Austin, Texas, whose stated goal is to become "the Norton Computing of the cloud computing era." They want to be the provider of all the things that should have come "in the box" with modern cloud applications like those from Google, Microsoft, and Salesforce.com, but didn't. They are well on their way, with tens of thousands of customers in 58 countries and just six full-time employees.

Charlie Wood, one of the founders, describes the journey:

> When Spanning got started, we did so with very little in the way of resources. There were two of us, and between us we had a couple of computers, a free hour or two here and there, and not much else. But we were able to use a broad array of free or cheap tools to help us get started. My business partner Larry Hendricks was in California and worked late nights while I was in Texas and worked early mornings, so we needed a way to work asynchronously. We found the solution we needed in the free version of Google Apps, which we used to communicate via email and collaborate on shared documents. We built our product on a stack of free software: LAMP on the servers, and Apple's free developer tools and an open-source framework called Cappuccino for the client. We built a mailing list of people interested in our product with Google's free FeedBurner system, and tracked visitors using Google Analytics. Once we started selling our product, we used PayPal to handle the transactions so we did not need to open a merchant account at a bank. None of the systems we used required more than $100 to get started.
>
> Free and cheap tools allowed us to build a business from scratch with essentially no up-front investment. But for that business to succeed, we needed promotion and distribution, and we got it. We were early participants in the Google Solutions Marketplace, which has grown to become the Google Apps Marketplace. It's essentially a catalog of third-party tools and services that extend Google's products, and it provides customers an easy way to find our products and rate the products they've used. Now that we're selling products into

> businesses, it also provides IT administrators at those companies a way to integrate our products directly into their Google Apps installations, which allows us to provide more valuable functionality. Google also invests in the Apps Marketplace ecosystem by providing online forums for marketplace participants to report problems and discuss best practices.
>
> Spanning is both a producer and heavy consumer of technology. The products we sell help individuals and companies work more efficiently. But without the vast array of technology available to us at little or no cost, we wouldn't have been able to create those products in the first place.

Entrepreneurs like Wood and plenty of others we discuss in Chapter 7 can work wonders with $100,000. Infrastructure and applications can be had on a pay-as-you-go basis with cloud computing options. There are open APIs to be leveraged from PayPal, Facebook, eBay, iTunes, Amazon, and so many others. Social networks allow for affordable viral marketing. There are ecosystems around Apple, Google, Amazon, and others that allow for pay-as-you-go selling costs.

Even the funding process has evolved. Phil Simon, author of *The Age of the Platform,* says, "Sites such as Kickstarter and IndieGoGo allow creative types to raise funding and awareness without very much effort. I used Kickstarter to raise more than $4,000 and $7,500 for my third and fourth books, respectively."

Simon uses a creative tier of incentives for his backers—as an example, if you "pledge $200 you'll get your name in the acknowledgments plus a signed physical copy of the book with an inscription. We'll also schedule a one-hour video chat to be recorded on Skype. You can ask me questions about whatever you like, up to and including how you can build a platform for your business."

Says Simon: "Other Kickstarter projects have generated 10 or 20 times as much funding."

Actually, the most successful Kickstarter funding has been for an iPod Nano watch kit that raised close to $1 million from over 13,000 backers—and the original goal was to raise only $15,000![4]

This makes Simon's point: "The days of waiting to be 'chosen' are coming to an end. Equipped with sufficient moxie and simple websites, no longer do creative types have to suffer in anonymity."

They cannot be guaranteed success, but they will learn to buy and sell technology, to switch-hit for very little. And do so while they are young.

Let's Not Underestimate—Switch-Hitting Is Not Easy

Whether you start early or later in life, switch-hitting is not easy. Ted Simmons, a coach for the San Diego Padres, has been quoted as saying, "I also have yet to find a person that completely, totally, unequivocally has bilateral symmetry. One side is always dominant" and "I've asked concert pianists if it helps them to be ambidextrous. They told me, 'No. It's just that the left hand is trained.' They don't switch pianos for left-handed people. But they switch guitars for left-handers. Jimi Hendrix was left-handed. He switched wires. But being ambidextrous doesn't help as a guitarist."[5]

Of course, there is need for more practice. Not favoring either side means working out more. Then there are the mood swings as one side slumps from time to time. Lance Berkman, considered one of the best switch-hitters ever (and member of the 2011 Major League Baseball champions, the St. Louis Cardinals) is on record as saying, "If I could do it all over again, I would not be a switch-hitter."[6]

Similarly, becoming a technology "buyor" is not easy. If you are in Indiana or in Indonesia, how do you acquire the DNA of Silicon Valley? On the other hand, how do you bring grocery chain frugality to a software company accustomed to 90 percent margins?

Take the example of Xerox, which many in Silicon Valley think made lots of others, including Apple and Pixar, wealthy by not commercializing what they innovated. "The multitude of technologies that Xerox threw away were later commercialized outside of Xerox's company, and enabled other companies to profit from some of these quite valuable ideas."[7]

Bob Warfield, a serial entrepreneur, has seen even technology-savvy companies stumble digital transitions. "Before home theaters evolved, high-end audiophile gear was largely analog. The analog engineers were in charge at those organizations. They had to bring in digital/computer guys to deal with the largely digital home theater medium. Problem is, when the analog guys are king, they think they can call the shots, make the decisions, and hire underlings to do the digital. As a consequence, their gear was unreliable and not up to the standards of newer players that had empowered the computer thinkers to call the shots on that gear."

He thinks as even less tech-savvy industries plan their own smart products, they risk underestimating the risk of their digital journey. "Computers are the only machines that can change their function. They are universal meta-machines. A phone without a computer is just a phone. A phone with a computer is a phone, and MP3 player, web browser, email client, calendar, and a thousand other things. A phone that is a computer can do new things. Aside from being universal meta-machines, computers remove friction. That's friction to change, to learning, and to growing. These are all massively disruptive to industries new to computing. Businesses that don't understand meta-machines have a terrible time succeeding in the digital world, and eventually all things turn digital."

The Need for a New Breed of Ambidextrous Technology Executives

Today over 15 percent of Major League Baseball hitters are switch-hitters, They are switch-hitters not just to honor Mickey Mantle, but because there are significant advantages to switch sides depending on whether a pitcher you are facing is right- or left-handed. Every sport has a finite number of spots in team rosters, so multitalented players are valued.

Where do you find the "switch-hitter" equivalent of technology executives?

An *InformationWeek* survey showed "most CIOs still haven't embraced this idea of IT being part of the product, though the percentage is growing. Thirty-four percent of the more than 200 IT leaders we surveyed say introducing 'new IT-led products and services for customers' is among the top three ways they'll innovate this year. Two years ago, just 18 percent considered that a priority."[8]

Bruce J. Rogow is a former head of research at Gartner and now runs IT Odyssey & Advisory, a private research and second-opinion practice. He annually visits with over 120 executives involved in managing IT. In an interview, Rogow says,

> For the past fifty years, IT has been focused on the continuous improvement and automation of business functions and processes. Despite their rhetoric, Chief Information Officers (CIOs) have top-down designed and delivered systems to users and customers on a one size fits all, common global solutions basis. The emphasis has been on lowest IT cost and business efficiency. The systems were directive in that they were provided on a take-it or leave-it basis.
>
> Since the summer of 2010, however, I have seen a massive shift in the role, expectations, focus, and activities of IT within enterprises. While there are still efficiencies, economies of scale, and incremental improvements to be gained for competitive advantage, the business is looking for three very different outcomes from IT.
>
> First, the focus shifts from supporting broad continuous improvements of business process to a game of business arbitrage. Business arbitrage involves being able to quickly identify and fill gaps between what competitors provide in a market and a customer need, price point, service offering, or expectation. It may be diverting an existing product to a new channel to rapidly fill a gap left open by competitors. These gaps open and close rapidly.
>
> Second, IT is expected to contribute business revenue through marketing, product enhancement or new service lines. This often moves IT from the back of the house to a customer facing role at the front of the house.
>
> Third, the focus is also on IT to dramatically enhance business yield. CEOs I have interviewed consistently lament that their biggest issue was inadequate yield from the elements and assets of their businesses. IT is being asked to increase the yield of product development efforts, revenue from the clients, value derived from the intellectual property, impact of the shared knowledge, and assets of the firm. This is not about making people more productive so head count can be reduced. It is about making the assets and people more effective and impactful.

Another Model—Bypass IT When It Comes to Product/Growth Initiatives

Some companies are starting to emulate the Silicon Valley model. Let the VP of R&D or Product Engineering develop products and let the CIO focus on internal technologies. As an example, Daimler, the leading auto and truck manufacturer, has more than 1,000 software engineers in their R&D group, compared with hundreds of others who support internal IT systems like SAP.

There is another good reason to think of R&D more than IT when it comes to new product development. Gina Staudacher, partner at the accounting firm of Baker Tilly, advises clients about R&D credits and how to structure such product development projects. She says:

> These significant technology investments are often eligible for the Federal Research Tax Credit and corresponding State R&D incentives. This gives companies a monetary payback via reduced federal or state income taxes. These research credits, although most often secured "after the fact," provide incremental cash benefits to an otherwise necessary business IT spend. Unfortunately, quite often these R&D incentives are misunderstood and not claimed. These incentives vary among states and foreign jurisdictions but remain relatively consistent in their purpose. These tax and cash incentive programs are fundamentally based on "product improvements." To the extent they involve a process or technology enhancement, those activities and costs can be defined as qualifying projects, eligible for these valuable incentives.
>
> The key is to understand the business purpose and underlying technology necessary to achieve the desired outcome. Ironically, in all of these R&D incentive programs, the investment that is being "subsidized" via the credit or incentive program is actually the exploration or development of a new technology or product or process.
>
> In evaluating whether a research credit is available, companies need to consider whether a technical advancement is being attempted, not necessarily achieved.
>
> When we work with clients to monetize tax credits globally, we utilize a "concept through final completion" approach to consider all potential project costs that may be eligible for R&D credit inclusion

from a global perspective. Hence, consulting fees incurred in a foreign country may not be eligible for inclusion in the U.S. R&D tax credit programs. Such costs and payments may, however, be eligible in the foreign jurisdiction wherein they were performed or where payment was made.

Enterprises Have Poor Track Record with Technology

No matter where the project resides, in IT or R&D, the enterprise track record with technology has been less than assuring. It would appear delays and overruns in technology projects are almost guaranteed. Embedding technology in products, in some ways, jeopardizes more than a failed IT project, which can mean painful write-offs. Failed smart products can tarnish brands, lead to customer dissatisfaction, and, worst case, even lead to product liability.

Take the example of the 2011 Ford Edge. *Consumer Reports,* a significant influence in auto consumer decisions, wrote this about its MyTouch console: "The bad news is that the MyTouch controls give a new meaning to the word 'unfriendly.' Most of the controls have lost their simple knobs and have been replaced with touch-sensitive buttons that give no tactile feedback. The system also has busy touch screens that force you to take your eyes off the road too much. All-in-all, it's an aggravating design."[9]

It extends way beyond Ford. As J. D. Powers reports, "A variety of issues led to the unexpectedly poor performance of all-new 2011 models—the first time since 2006 that newly-launched products haven't improved in quality compared to the vehicles they replaced." The most notable were engine/transmission problems, according to Powers Vice President David Sargent. But there were also significant issues with the latest audio/entertainment and navigation systems.[10]

Crossover Executives

Enterprises are finding that they need "crossover" executives who can bring a blend of skills—vendors are hiring talent from user organizations; user organizations are hiring talent from vendors and startups.

A good example comes from the continuing competition between Walmart, the largest retailer in the world, and Amazon, the leading online retailer. In 1999, while Amazon was still a fledging player, Walmart sued it for hiring several of its technology and logistics executives. While the suit was settled, it reflected Amazon's building of a strong team across various organizational areas. The previous year, it had acquired a startup called Junglee, which had pioneered Internet comparison shopping. That acquisition brought it two of the co-founders, Venky Harinarayan and Anand Rajaraman, who were influential in its formative years.

Now, fast-forward to 2011 when Walmart, this time, bought Kosmix, another company that Harinarayan and Rajaraman had started. The two executives are now running @Walmartlabs "to speed with innovations such as smartphone payment technology, mobile shopping applications, and Twitter-influenced product selection for stores" to give Walmart a chance to narrow the widening lead Amazon has in online retailing.

Examples of other crossover executives—Tony Scott, CIO at Microsoft, went there after stints at Disney and GM, and in the other direction, Vijay Ravindran, the Chief Digital Officer at the *Washington Post,* who did stints at Amazon and other startups. Scott and Ravindran write about their crossover experiences in the next section.

Beyond Ambidexterity

Michael Cote, former analyst at Redmonk (since this interview, he has moved to a corporate strategy role at Dell), raises other issues regarding the state of enterprise technology both in corporate IT and at technology vendors. He says:

> Enterprises had their chance to be the flagship of IT, as did the military and the scientific community before them. All of those leaders did really well and gave us relational databases, GPS, processors and memory that perform beyond our wildest dreams. All of these things were good, but then business got wrapped up in developing software that was not focused on being usable, but intended to be compliant,

safe, and affordable above all else. With rare exception, business no longer looked at computers as a way to differentiate and drive business. Their speed at adopting new technologies belied any attempts. Businesses were dragged into e-commerce and banks took forever to do rational, online banking.

The consumer space is where interesting technology innovation happens now, and taking a consumer mind-set is what works when applying technologies. Consumers are not fixated on risk management. If I use Facebook, am I opening myself to privacy risks? If I sign up with the Sony Playstation Network, what's the risk/reward for potential credit card leaks relative to the enjoyment? If I use this Android phone, what happens if Oracle wins their lawsuit against Google? How did my new Galaxy Tab know about all the wi-fi networks I had connected to on other Android devices?

Instead consumers look at using computers to deliver functionality they need, not as part of some grand, *management-chain* plan.

Consumer tech is focused on the user, not the "stakeholder." Enterprise tech has come to care only about the second.

Steve Jobs famously stood at the crossroads of liberal arts and technology, to summarize the breadth of skills that makes Apple so successful. That is, however, deceptively simplistic.

To get to Rogow's "business arbitrage" and to Cote's "consumer mindset," a technology group needs to be far broader—focused on the 12 attributes we cover in Part II. They include elegance in product design, global orientation in customer nuance analysis and supplier diversity, ability to stand out in social media and in physical settings, dramatically increased "clockspeed," and many more facets. The conversations in many technology communities these days are related to cloud computing, virtualization, "big data" analytics, service-oriented architectures, and security concerns around LulzSec. Clearly, every technology group needs a wide variety of technical skills, either on staff or contracted, but even that is not wide enough.

To continue with the switch-hitter analogy, players cannot just be good hitters; they are much more valuable if they are also good fielders, they are quick enough to steal bases, and they possess other athletic abilities. Similarly, a soccer player who can kick well with both feet is also expected to play good defense. The elite are multidimensional.

Conclusion

Just as in baseball, where only 15 percent of players are switch-hitters, the ambidextrous technologist who is good at both technology production and efficient at technology consumption is an elusive species. We need them and we need many of the crossover types who bring experiences from both technology vendor and user organizations. We present next perspectives of Scott at Microsoft and Ravindran at the *Washington Post,* both examples of crossover executives.

Guest Columns: Crossover Executive Perspectives

Perspective 1: Tony Scott (CIO, Microsoft)

Tony Scott joined Microsoft Corp. in February 2008 as corporate vice president and chief information officer (CIO). Under Scott's leadership, Microsoft IT is responsible for security, infrastructure, messaging, and business applications for all of Microsoft, including support of the product groups, the corporate business groups, and the global sales and marketing organization.

Before joining Microsoft, Scott was the senior vice president and chief information officer for The Walt Disney Co., where he led planning, implementation, and operations of Disney IT systems and infrastructure across the company. He also held the position of chief technology officer, Information Systems and Services, at General Motors Corp. (GM), where he was responsible for defining the information technology computing and telecommunications strategy, architecture, and standards across all of GM's businesses globally.

Today, every business is increasingly becoming a "digital" business in the sense that many or all of the critical activities that business engages in are digital in nature. Starting with product development and continuing through marketing, selling, the buying experience, and product support (and sometimes the product itself), the common thread is a complex "circuit board" of digital information flow and technology-enabled business decisionmaking. For consumers, it is crystal clear when an organization has embraced and exploited the digital terrain and when it has not. For the CIO and the IT organization, it is of the utmost importance and priority to get this done right and to keep up with the pace of change.

One of the most important digitization changes that is taking place in many businesses is the blurring of what were once very distinct and separate internal functions. A couple of great examples come to mind from my own experience. At Disney, as our creative process teams began to use digital technology to create films, TV shows, and consumer products, the support teams that used to process and edit film and magnetic tape, or physically move it around in bulky containers, began to use digital technology to record, store, edit, and transport content. Often, the servers, network gear, and storage devices that were used to support these activities were the same kind of gear that the IT department was buying, and sometimes lived in the same data center. We recognized that there were some big differences in terms of the work being done on these servers, but the underlying hardware technology was almost exactly the same. And the management tools and disciplines that are required to run these environments didn't vary much, despite the differences in workload. As time has gone on, much of what was once differentiated in terms of infrastructure has become commoditized, and the focus has shifted from hardware of various kinds to software.

At Microsoft, a similar pattern has emerged. We once shipped the vast majority of our products in physical form. Packaging, physical distribution, inventory management and returns have gone (or are quickly going) the way of the floppy (who needs them anymore!). We now sell, distribute, and support much of our software in digital form—bits over the wire. Indeed, some significant part of our business is software as a service (SAAS), where our customers don't actually own or license anything! The underlying technology that supported the physical oriented world has given way to content distribution networks and digital contact centers that use a ton of information technology as the base platform.

All of the above present new challenges and opportunities for the IT department, as we are all being asked to manage exponentially increasing quantities of servers, storage, and network bandwidth, along with a smorgasbord of consumer devices, and at the same time deliver high quality real-time information and analytics/business intelligence platforms.

It is quite clear to me that future CIOs and the IT organizations they lead will have to stretch to learn new skills to adequately support the digital brand experience that consumers and internal customers will expect, while still protecting important intellectual property and other information assets of the organization. I think it is an exciting time to be in our business, and every day I am energized by the creativity and imagination I see being applied to some really hard problems across dozens of industries and professions. And I am encouraged by the healthy levels of both collaboration and competition our industry drives. I hope that you'll find the journey as interesting and exciting as I have, and that you'll appreciate the progress we make together.

Perspective 2: Vijay Ravindran (Chief Digital Officer, The Washington Post Co.)

Vijay Ravindran joined The Washington Post Company as senior vice president and chief digital officer in February 2009. In his role, Ravindran focuses on digital news product development. Ravindran founded and leads WaPo Labs, which develops experimental news products, including the personalized news aggregation site Trove. Ravindran also serves in various leadership roles at SocialCode (the leading Facebook-dedicated advertising agency), The Washington Post's *online initiatives, Slate, Avenue100 Solutions (a leading analytics-based performance marketing company), and Ongo (a Company investment).*

Previously, as chief technology officer of Catalist LLC, a start-up political technology company that built a national voter database and data mining tools for political campaigns, Ravindran led all the technology aspects of developing the company's software products. He joined Catalist at its inception in late 2005. During the 2008 election cycle, Catalist clients included the Obama for America and Hillary Clinton presidential campaigns.

Prior to Catalist, Ravindran was a technology director at Amazon.com. From 2003 to 2005, he led the ordering services group; the department was responsible for consumer purchasing on all Amazon properties. From 2001 to 2003, he built and led the teams that owned the core order-processing and identity services for Amazon and its partners.

In the late summer of 2008, I was approached by an Amazon director, my former employer, about my possible interest in a unique position being created by Don Graham, the CEO and Chairman of the Washington Post Co. The position would be focused on digital news, specifically of innovating beyond the quarter-to-quarter improvement all operating businesses hope to make, but instead on a "big swing" looking out five years and beyond, and seeing if we could "skip a generation." So began the recruitment and my eventual hiring as Chief Digital Officer.

There are some reasons to stay away from "newspaper" companies. Their main business is deeply challenged, and while there are audience riches online, the same cannot be said of the current advertising models. Original content around "serious" news does not, at least at face value, seem like a good business to be in for this coming decade.

So why take the leap? I was leaving a startup, Catalist, that I had started at on day 1, built and shaped the product, hired the technical team, and been through the war of a presidential election. My time at Amazon prior to Catalist had given me the cushion to take chances and the skills and demeanor to constantly look for big problems to solve. This was a big challenge. Given my time in Democratic politics, my respect for the *Washington Post* newsroom was so high that to play a role in helping transition the essential good qualities of that newsroom to a new age of information delivery was something I could really believe in.

That by itself wasn't enough; I needed a role that would allow me to be both influential but not be bogged down. I wanted to be a game changer, but given the latitude to not always be focused on a P&L and free cash flow in the short or even medium term. And to be at a company where that sort of role could be supported and respected. Most newspaper companies do not have the balance sheets to support that financially. The Washington Post Company was ideal because it was financially strong, thanks to Kaplan and CableOne. Don and I developed a common shared vision of how innovation could work from outside the *Washington Post* operating unit while still being relevant and respected within.

My job description, "to develop next generation products, strategies, and investment to allow the Washington Post to skip a generation" was definitely vague enough to be considered flexible. When I arrived in February 2009 (I needed time off from the election to recover), I set out to understand the business, culture, and personalities. I was personally more interested in the investment side, and I had originally envisioned leaving Catalist for a venture capital firm to either be an executive-in-residence or join a firm itself. Quickly, though, I realized that the value of a technical team outside of the operating unit, built in the vision of the type of team I worked with at Amazon, could be the most effective use of capital to make a difference.

While I set out to begin hiring what would eventually become the WaPo Labs team, I also was given corporate oversight of one of the smaller businesses within the portfolio, CourseAdvisor, an education lead-generation company that has since changed its name to Avenue100. Thanks to the CEO of the Washington Post, Katharine Weymouth, I also had fairly open access to the Post's VP group and became a de facto member.

The biggest impact I had in the first year, and continue to have, is to reshape the image that technical leaders have to senior management both at Corporate and the operating units. They had only thought of technology as "IT" and the technical organizations were fairly subservient organizations. By Don's creation of my position, I've been able to showcase the level of product and business thinking that can turbocharge having a deep technical background. The kind of technology executive who is at Amazon and its peers had finally made it to a company that had never seen anything like it before.

This role demands tremendous breadth. I currently divide my time across a variety of initiatives and roles:

- The WaPo Labs team continues to grow and the team launched Trove, a personalized news aggregation site with associated apps. The team is mainly focused on growing Trove as consumer destination while componentizing aspects of the personalization technology to offer back to Washington Post and Slate. Trove offers cutting-edge personalization of news using users' Facebook personas to help create a unique news reading experience.
- Using technology originally developed at Avenue100, we've incubated SocialCode, a Facebook advertising agency that leverages Facebook's self-service advertising platform but targets big brands and their agencies as clients. The business has gone from not existing 18 months ago to being a real company with dozens of employees working across many dozens of clients.
- I work closely with the leadership of Slate and Washington Post on new endeavors and selected business development. Both sites are undergoing major software infrastructure upgrades, including their core content management processes. In the case of the Washington Post, the entire online business was managed as a separate company until 2008, and the integration has been a complex process of people and infrastructure.

- I serve on an advisory group for Kaplan Test Prep's innovation efforts. Kaplan's Test Prep business faces a completely different set of challenges than a decade ago. While my current role is quite small, there are quite a few legacy parallels to the challenges faced by the *Washington Post*. For instance, just like *Washington Post* faces niche content challengers across local news and politics, every specific test prep area is a battleground for Kaplan with new startups.
- I help TWPC make several angel and Series A investments in technology startups, including those in Ongo and Fab.

While I do not have the luxury of 100 percent focus on a product like a startup CEO, I feel like the opportunity exists to do one part "pure innovation" and one part weaving in a set of assets in new innovative ways, and that the sum of this can be much greater than the parts if executed. The path I've taken, while fairly unique today, is something that I think carries great promise for leaders from the Valley and other high technology organizations. I have been able to go from being a one of many directors and VPs at Amazon to making a more dramatic, higher-leveraged impact at an organization that has lots of opportunity to be successful in this next decade across a variety of business lines.

Part II

Key Attributes for the New Technology Elite: Three Es, Three Ms, Three Ps, and Three Ss

Chapter 6

Elegant: In a World of Flashing 12s

An Amazon subsidiary called Lab126, which designed the Kindle ebook reader, says on its website: "We want the devices we design and engineer to disappear as you read. . . ."

Barnes & Noble's response with Nook: "Books don't have buttons . . . so we felt that was not only an authentic place to be but also great competitively against the Kindle"—which has a keyboard.[1]

Microsoft says in a product announcement: "With Kinect, technology evaporates, letting the natural magic in all of us shine."

Bose says about its VideoWave Entertainment System: "No clutter, no confusion."

Not just in reading and entertainment, we are seeing technology simplification even in more complex areas. Siemens Healthcare, in introducing its MAMMOMAT Inspiration Mammography System, emphasized its light panel, which provides a warmer environment for the patient by illuminating soft, pastel colors.

The De Dietrich DTiM1000C 90CM Induction Hob features multiple inductors beneath the ceramic glass surface. Should you move any of the pans, the hob's automatic pan detectors ensure that the temperature of the pan remains constant, wherever it is moved.[2]

Technology "disappearing" and "evaporating" . . . welcome to the new age of minimalism even as technology proliferates around us and our old VCRs continue to flash 12s.

To emphasize that minimalism, as we described in Chapter 1, we have folks like Kelly Sutton who are aggressively trading digital technology for physical assets: "I think cutting down on physical commodities in general might be a trend of my generation—cutting down on physical commodities that can be replaced by digital counterparts will be a fact."[3]

Product and service design are a huge differentiator in this new world of minimalism even as technology proliferates around us.

Human-Centered Design

Mickey McManus says he has the best job in the world, and the more you listen to him, the more you have to agree. As CEO of Maya Design Inc., he has helped countless clients across industries think about product and process design. Maya has been named in the top "best small companies in America to work for" by *Entrepreneur*, *Inc.*, and other magazines. Its staff includes "cognitive psychologists, ethnographers, computer scientists, mathematicians, visual and industrial designers, brick and mortar architects, game designers, and filmmakers."[4]

The Maya site calls McManus "Chief Mad Scientist," and you get a sense of that if you watch a presentation he did at TED where he talks about containerization, liquid currency, and the possibilities of a trillion-node world and what nature's design patterns can teach us about computing.[5]

Don't let that fool you—his firm is about simplification. "On one end is the customer who is cautious, almost scared to use a product. At the other end of the spectrum is smug. We help iterate products to a point where most users can feel smug."

His basic philosophy—technology is evolving way faster than humans are:

> The most complex technology in many towns a century ago was a printing press. Look at what each of us, not the whole town, has access to today. We make more transistors today than grains of rice and we make them cheaper. And they end up in all kinds of "smart" products you and I use. Our research projects a trillion devices in a few years.
>
> But our brains have not evolved that dramatically in the last century. We like to say younger consumers are digital natives and will cope better. Studies have shown them to not be very good at multitasking either.
>
> We struggle with numbers in billions. Now think about dealing with trillions. Nature has, on the other hand, plenty of trillion cellular examples. Humans have been dealing with information, for say 6,000 years. Nature has been dealing with information for 3 billion years.

Then he launches into the talk he gave at TED. Make him pause and he goes, "We have plenty of technologists. We need more 'human scientists.' "

He describes Maya's work with SoloHealth, a company started by Bart Foster, with years of prior experience at CIBA Vision. Its introductory product was EyeSite, a self-service vision-testing kiosk—the kind you see at stores like Walmart. The initial tests with consumers showed they were taking as long as 15 minutes and many were giving up. Maya helped observe, interview, and videotape users at the pilot site. "We got it to the point where 90 percent of people sat through it and would recommend it to a friend. We got the test down to five minutes. There was a 17 percent increase in sales in eyewear at stores which had the kiosk." Just six months after that project, EyeSite was the talk of the April 2008 KioskCom Self-Service Expo and the Digital Signage Show. It garnered three Awards of Excellence: "Best Healthcare Deployment," "Best New Innovation in a Kiosk Deployment," and "Best in Show."

"To achieve ease of use, you need to forget technology for a while and get inside the heads of users."

Google's Doodles

Google's simple search box interface and its discipline of no more than 28 words on its home page set the standard for minimalism in technology.[6]

Indeed, to add the word Privacy (which linked to a policy statement) at the bottom of its page, it replaced its own name to not violate that constraint.

Even within that constraint, Google has managed to blaze a trail with its doodles.

"When doodles were first created, nobody had anticipated how popular and integral they would become to the Google search experience. Nowadays, many users excitedly anticipate the release of each new doodle and some even collect them!" says its website.[7]

The doodle team has created over 300 doodles for Google.com in the United States and over 700 have been designed internationally.

In August 2011, a doodle celebrated Lucille Ball on what would have been her hundredth birthday. The doodle was a graphic of a vintage 1950s TV set, and you could change channels to seven favorite scenes from *I Love Lucy,* including the hilarious one where Lucy and Ethel try to stay ahead of the conveyor belt in the chocolate factory.

Some of the best and most challenging doodle ideas come from users.

> We got an email from a French astronomer who said we would be able to see the planet Venus crossing the face of the sun. We asked him when it was happening, he said "tomorrow." We told him we didn't know if we could do it on such short notice. He replied, "well, the next pair of transits is 122 years from now." So in less than 24 hours we had a design up all around the world. People could see a black dot that would slowly move around the sun.[8]

On Valentine's Day in 2007, the doodle caused major consternation, says Dennis Hwang, then Google's chief doodler:

> It was a design decision to have the strawberry represent the G and the stem represent the L. But I guess the stem was too short and people didn't make the connection. I was getting phone calls in the middle of the night after it went up saying we forgot the L. So I knew it was going to be a long day.
>
> So many theories were going around that we had to put out an official company blog post to users telling them that the stem was the L.[9]

Who says minimalism has to be boring?

Social Design

History tells us that Alexandria, Egypt, once had the largest library in the world, which was tragically burned to the ground. In its place today stands the Bibliotheca Alexandrina, a majestic structure designed by the Norwegian architecture firm Snøhetta. The Oslo Opera House, which won the European Union award for contemporary architecture in 2009, was also designed by Snøhetta. Its sloping roof slides into the water's edge, and visitors can walk freely over the building and the roof.[10] Snøhetta has also been selected as the chief designer to transform New York's Times Square to improve "the pedestrian experience in the plazas as well as the infrastructure for the various events held in Times Square throughout the year."[11]

The firm's philosophy is "Architecture cannot be contained by rules of order. Instead it must accommodate the restless mind of human society. It must accept associations developed by larger number of characters beyond the discipline of the architect."[12]

Fast Company named Snøhetta one of its 50 most innovative companies in 2011 and wrote that its designs "consider a structure's social experience—how the user enters, passes through, and lives in a building—to be as important as its form."[13]

3D Printing and Crowdsourced Design

Says *Fortune*:

> [3D printing] could have profound implications for manufacturing and design. Here's why: The blueprints or files for 3D printing are digital and, as a result, can potentially be edited by anyone—if the designer permits.[14]

Mike Prosceno, who works for the software company SAP and is an avid fisherman, marvels, "I was on a fishing forum where they were discussing how they could use 3D printing to produce lures in the exact color combination based on what the fish were biting."

Prosceno and consumers in all walks of life may soon get a chance to help design their products. How about a plane with no middle seats?

How about a car that stalls when the driver raises the stereo volumes to obnoxious levels?

New Interfaces for Cars and Cocktails

The QWERTY keyboard was originally designed in the 1860s for typewriters we don't use anymore. It was a clever design to prevent jamming from common combinations of alphabets. You would think that long ago someone would have solved the jamming problems rather than persist with the dysfunctional layout. Didn't happen, and so the majority of the world still interacts with computers with technology that annoys our eyes, tires our hands, and mostly ignores our other senses. Particularly gnawing is that software vendors go through waves of "next-generation" user interface projects but keep wasting billions because they don't seriously question the role of the keyboard.

One industry that has been exploring a wide range of visual, audio, haptic, and other user interfaces is the auto industry. In the BMW 7 Series, "The nerve center is the iDrive system, which has four direct-entry buttons that provide the tactile interface to the driver. The car's steering wheel has a vibration device to alert the driver. The 10.25-inch, high-resolution display provides the visual interface. Its rear screens provide passengers with a vibration-resistant visual interface. The sophisticated Logic7 speaker system and Bluetooth-enabled microphone provide the audio interface."[15]

Mercedes is looking to evolve user interfaces even more with its Cam-Touch-Pad HMI (Human-Machine Interface). "The system utilizes a center console-mounted trackpad similar to the one on your laptop. The difference is that the pad is translucent, allowing a camera mounted behind the pad to track your finger movements. A transparent image of your fingers appears on the dash display, allowing you to see your fingertips glide across the screen while you manipulate the various controls. The image never covers up the virtual buttons and the pad allows you to swipe, push, pinch, turn, and rotate everything from maps to climate control settings. Mercedes also plans to introduce 'Gloria,' who is 'a digital avatar' that's displayed on its COMAND (its vehicle command center) screen to take voice commands ranging from navigation instructions to point-of-interest searches."[16]

Of course, Gloria, though impressive, is no match for IBM's Watson. While Watson dazzled everyone by beating experts at the game of *Jeopardy!*, the real gem from that exercise was the computer's ability to process natural language. "I think Watson has the potential to transform the way people interact with computers," said Jennifer Chu-Carroll, an IBM researcher working on the project. "Watson is a significant step, allowing people to interact with a computer as they would a human being. Watson doesn't give you a list of documents to go through but gives the user an answer." Just a few months later, Apple introduced Siri, its "personal assistant" in its iPhone 4S, which understands many commands in conversational English, French, and German.

Why even bother to talk? There is the brain-machine interface like the NeuroSky technology in the Mattel Mindflex Duel game described in Chapter 1. If you wanted to see a really head-turning application, though you had to be at FutureEverything festival in Manchester, UK, in May 2011, Absolut Vodka used the NeuroSky technology to show cocktail drinkers a visual representation of how their brain is enjoying their drink.

Keyboards? Definitely a technology that needs to evaporate!

The Most Valuable Executive at Apple?

No question it was Steve Jobs, right? Well, maybe not.

The London *Daily Mail* asks, "Who is the most valuable Englishman on earth? Wayne Rooney? Colin Firth? Neither gets near Jonathan Ive, the boy from Chingford who is now senior vice president of industrial design at Apple. Ive has given style to a family of machines that has changed the way the world thinks. Transient, global, instantaneous, intelligent, wireless connectivity is a bigger idea than the French Revolution. More than any other individual, Ive has decided what this idea should look like. And it looks beautiful, desirable."[17]

Ive has been quoted as saying, "One of the hallmarks of the (Apple design) team I think is this sense of looking to be wrong. It's the inquisitiveness, the sense of exploration. It's about being excited to be wrong because then you've discovered something new."[18]

Wow, it must be so right to be so wrong so often and design all those Macs, iPods, iPhones, and iPads!

Of course, Jobs also had the chance before his return to Apple to work with John Lasseter, the creative genius at Pixar. Pixar is the company that has mined animation into 26 Oscars, seven Golden Globes, three Grammys, and over $6 billion in worldwide sales. The story of Pixar was long and tortured before its success. Says Lasseter of Jobs, "He waited over 10 years, and supported us all that time, before we had a hit with *Toy Story*. In today's business market, where everyone wants an immediate return on their investment, that's sort of amazing."[19]

Pixar was all about cutting-edge technology (by the standards in the 1990s and since). Woody (the main character in *Toy Story*) "had over 723 motion controls to animate his actions. In his mouth alone, there were 58 such controls. That type of detail required an incredible amount of computing power. It took hours for 117 Sun and SGI servers to render each frame, and the film had hundreds of thousands of frames."[20]

A *Los Angeles Times* story on Lasseter says, "He loves telling stories, that's when he's happiest"; "The corporate stuff (now at Disney, which acquired Pixar) he's incredibly good at, but it's not fun for him. He's a big kid, and his playbox is Pixar."[21]

As Jobs told Charlie Rose in an interview in 1996, well-made movies, unlike most technology, last for decades.[22] He invoked the enduring popularity of *Snow White*, made by Disney six decades earlier.

That is a good measure of elegance. Let the technology disappear into the background and allow the consumer to enjoy the experience even years later. Even better, allow the consumer to be smug about it.

Where Do You Find the Innovators?

Robert Brunner, like Ive, has an impressive design track record. He is credited with Apple's Powerbook and Newton and Amazon's first Kindle. He is now helping Barnes & Noble (redesigning the Nook) and Polaroid, among other clients at the firm Ammunition.

Where do you find the Brunners and the Ives?

"Not in business or engineering schools," says Bruce Bendix, Director of Growth Strategies at Baker Tilly. He continues:

> Business schools do a poor job of creating innovators. Emphasis on the analysis of *what is* rather than consideration of *what could be* keeps business schools firmly entrenched in the present instead of looking toward the future. The case method I was taught at Harvard Business School reinforces decision options already presented to you rather than creating new alternatives. Financial tools with their bias toward short term, low-risk, low cost options over longer-term, bolder strategies are notorious for stifling innovation. In part this is because the value of innovation is often difficult to quantify. As Albert Einstein famously said, "Not everything that can be counted counts, and not everything that counts can be counted."
>
> For me, new-generation "d-schools" are producing business professionals well trained in the discipline of innovation. An example: Stanford's d-school founded by David Kelly brings an interdisciplinary approach to innovation by drawing upon students from all the professional schools to address challenging problems. (Kelly is also founder of the design firm Ideo, which takes credit for a wide range of innovative products like Steelcase's Node Chair, which allows classrooms to transform from static to mobile, and the self-warming FeedMe baby bottles for Yoomi).
>
> My own experience has been with the IIT Institute of Design (ID) in Chicago. Considered the leading graduate design school in the country, it teaches design methods ranging from user observation to mine deep customer insights, contextual research to understand opportunity spaces, rigorous analysis of qualitative data, structured ideation, business concept generation, visualization and rapid prototyping, and leadership aimed at fostering cultures of innovation in organizations. The school sees itself as the most business-minded of the leading design schools, with its students going into roles in strategy and planning in a number of leading global corporations. I teach a class in Business Model Design at ID which was oversubscribed last quarter.

Conclusion

We are seeing a revolution in design in devices, in our software, in our architecture. If you aspire to be one of the technology elite, you have to put industrial design high on your agenda. In the next case study, we see how Virgin America, using good design and technology, is redefining the airline industry. This is an industry that in some studies scores lower in customer satisfaction than the IRS.

Case Study: Virgin America—Redefining Elegance in Flying

"With SaaS and cloud computing we are seeing IT move to an on-demand model. Virgin America has similarly brought on-demand to the airline industry."

Brett Billick, Director of Customer Loyalty Programs at Virgin America, is explaining some of the philosophy of the airline—including the subtle one of calling customers "guests."

"From our 2007 launch, our mission has been to harness new technology and design to reinvent the flying experience and give guests more comfort, choice, and control," says David Cush, president and CEO.

Yes, 2007 launch—this is the San Francisco-based airline and a completely separate company from Virgin Atlantic, the international carrier that has been in business much longer. The Virgin Group is a minority investor in Virgin America. They are both part of Sir Richard Branson's Virgin family.

On-Demand in the Airline World

Billick expands on the On-demand theme:

> On most airlines, the food and drink trolley comes along when it suits the flight crew. On Virgin America, you order your meal whenever you want during the flight and as often as you like.
>
> Upgrades—most other airlines have cut-off times, even days before your flight. On Virgin America if you feel like upgrading while on the flight you can via the airline's state-of-the-art entertainment system, Red.
>
> You want to work on the flight? The airline has wi-fi onboard every one of our planes. There's no guessing whether a particular flight has wi-fi or not and there's a standard power outlet near every seat.
>
> You want to be entertained? We have the largest entertainment library in the U.S. skies—not just one movie shown when you may want to nap.

You want to nap? We have one-of-a-kind moodlighting, which transitions through 12 shades that adapt to outside light that is incredibly soothing and not only helps to relax guests, but gently wakes them on a red-eye flight.

You want to shop? Most airlines have a paper SkyMall catalog and you hopefully remember to call when you land. We offer on-demand digital shopping via the Red platform at every seat.

You're curious about where you are? No need to pass on a request to the captain or wait for him to announce the plane is passing over the Grand Canyon. We have interactive, terrain-view Google Maps on every seat.

You want to use your Elevate program frequent flyer points? We have no blackout dates. If there's a seat, you can book it any time.

Continues Billick:

We also have custom designed, deeper, and more comfortable leather seats throughout our cabin. The Virgin America experience was designed to be like no other, marrying stylish design and innovative technology. The airline provides an upscale flight for less and gives guests control over their in-flight experience.

That was a core design principle when we launched our airline. Virgin America is a California-based airline on a mission to make flying enjoyable again—with brand-new planes, attractive fares, topnotch service, and a host of innovative amenities like wi-fi and the Red platform.

Red, a Linux-based in-flight entertainment system, is accessible from a nine-inch touch screen on every seatback and a Qwerty-keyboard handset at every seat.

Flying Redefined

Wired magazine has called their planes "a multimillion-dollar iPod. That flies." When you analyze the nuances, you marvel at the improvements it brings to flying:

The wi-fi is via Aircell's GoGo In-flight Internet described in Chapter 5. The service leverages Aircell's experience with airphone technology onboard flights in the 1990s. It uses air-to-ground frequencies—a big difference from the satellite services that other airlines are trying out. The plane transmits via underbelly blade antennae to over 90 cell towers that can be accessed around the United States and up to about 100 miles over international waters. Each plane has three wireless access points, and even if everyone on the flight uses the wi-fi service, the speeds are acceptable. On the day Virgin America became the first airline to have fleetwide wi-fi, the airline did a Skype chat call with Oprah Winfrey for her "Where in the Skype Are You" show. The Red system also allows a seat-to-seat digital chat capability so you can converse with a colleague a few rows away.

The shopping is via a special section of the SkyMall online store. You can buy a Tag Heuer Aquaracer Watch or you can drill into the Apple, Best Buy, Target, Barnes & Noble, and many other online stores. Each purchase rewards you with Elevate frequent-flyer points.

The enhanced Google Maps feature shows terrain views with eight levels of zoom functionality, so you can see the topography over which you are flying. Jesse Friedman, product marketing manager for Google Maps, has been quoted as saying, "With fresh data presented in our beautiful terrain view, this update improves travelers' ability and desire to track their flight progression as it happens. Flying is simply more fun when you can explore what's below."[23]

The dining and entertainment is supported with an "open tab" functionality. Guests can swipe their credit card just once per flight and order food, cocktails, movies, and more—and keep the tab going.

The entertainment includes live satellite television lineup via DISH Network, prerecorded television channels, and a 25+ film library at every seat. Alternatively, you can enjoy the 3,000+ MP3 music library and even create playlists in-flight. Need more choices? You can play a variety of videogames, including cult favorites like Doom.

Virgin America is known for its award-winning cuisine. The airline refreshes its offerings quarterly in order to provide travelers with seasonal and locally sourced food options. It offers food pairings like a Pacific brie and turkey sandwich bundled with PopChips and Peach White Honest Tea. Or you can order à la carte. The first-class menu is even more elaborate and has included entrees like a Beet Lavash Wrap with Gazpacho or Goat Cheese Tortelloni.

The drink choices are just as wide, with specialty cocktails named Funkin Margarita, Drunken Parrot, and Moonmosa. Virgin calls them "Cocktails with Altitude." Flight crew, by the way, see your food and drink orders on a tablet in the pantry of the plane where they prepare your serving and bring it to you on a tray.

Slim-line Recaro Aircraft Seating with a full-size headrest provides unusually comfortable airline ergonomics. Some of the nice features: 32-inch general seat pitch, five-inch average seat recline, hidden underfloor electronics boxes providing all seats an unobstructed foot well; a seat structure with greater knee clearance than traditional seats at the same pitch; luxurious black leather, soft upholstery, and high-gloss white plastics; headrests with adjustable wings as well as height and tilt adjustment; enhanced lumbar support, and contoured seat-pan structures unlike the metal bars that can be felt through traditional airline seats.

And that is seating in coach class! First Class seating offers 13 inches of electronically controlled recline with extended leg and footrests and other nice touches you typically see only on international flights. In another nice touch, cabin lavatory locations (behind the aft doors) mean that no passenger is seated adjacent to a restroom.

The planes themselves are new—fuel-efficient Airbus 320s. Virgin has announced plans to move to newer LEAP engines. "Designed in collaboration between CFM International, Snecma, and GE, the new LEAP engines improve fuel efficiency by 15 percent. Aside from fuel efficiency, the LEAP engine also produces 15 percent less CO_2 emissions than other planes and it reduces nitrogen oxide emissions by 50 percent. The LEAP engine also cuts engine noise by 15 decibels, which is like "reducing a jack hammer to an alarm clock."[24]

The planes are already carbon-conscious. Virgin was named "Most Eco-Friendly Airline" in May 2010 by the SmarterTravel Editors Choice Awards. In addition, Virgin allows its guests the option to pay a voluntary fee when booking their ticket, which will go toward supporting carbon-offset projects.

Billick continues:

> Technology has allowed us to leverage social tools effectively. In fact, we reward guests for virtual check-in via Foursquare and Facebook Places. We leverage our social media channels given that we have

> a number of very passionate and loyal advocates online. As the only airline based in Silicon Valley, our flyers tend to live online as well—and it's a great way for us to drive awareness. We were also the first airline to offer fleetwide wi-fi and plugs near every seat—so we have a real-time link to our guests via social media channels. We've also launched some interesting partnerships with some of our neighbors like Twitter, Google, and Loopt, and our partnership with Groupon—the first-ever Groupon for air fare—was very successful. Social media can be a real authentic link to our guests—it can help us drive awareness and trial as a young airline, identify positive and negative issues as they unfold, and increasingly we see that it can drive sales when done in a compelling way.

Loyalty Program 2.0

Billick again:

> Technology has also allowed us to create a very innovative frequent flyer program.
>
> To start with, it is simple—which in itself is innovative. You earn 5 points for every dollar spent, and free flights start at as little as 2,500 points. There are no blackouts, you can redeem for any unsold seat, and you can redeem for one way trips.

Billick continues:

> Or you can take advantage of some unique rewards like a party with Sir Richard Branson, or even a trip on Virgin Galactic, the group's planned private space venture.

Tibco's Loyalty Lab technology powers the Elevate program. Guests see dynamic pricing for various trip combinations in Elevate points as they would if they were paying for the fare. Loyalty Lab has made adding affinity partners much easier. Virgin America has over 200 partners that allow guests to earn points. These include credit card, hotel, and rental car companies, and other airlines. (Loyalty Lab is attuned to creative loyalty promotions. In one it helped Zynga award Farmville currency for General Mills spinach purchases.)

Loyalty Lab also provides Virgin analytics such as guest flight and award patterns, demographic and psychographic data, and partner activity.

"We've been very pleased with our partnership with Tibco's Loyalty Lab and the flexibility it has provided our Elevate program. It's extremely important for our guests to utilize Elevate redemptions as seamlessly as booking a regular flight—and Tibco Loyalty Lab's powerful technology has made it simple," says Billick.

It is pretty impressive what Virgin America has accomplished in four years. It now serves 15 airports in the United States and Mexico. But importantly, it has shown "the art of the possible" in an industry that *USA Today* said "scored lower than the Internal Revenue Service in customer satisfaction" around the time Virgin America was started.

Of course, it has earned plenty of kudos. *Travel and Leisure* magazine has named it "Best Domestic Airline" four years in a row—2008 to 2011. The real satisfaction comes from delighted customers like Karen Auby, senior manager of public relations at Plantronics, Inc. "I flew Virgin America for the first time in 2008 and have been a loyal customer since. I actually drive 20 minutes out of my way to SFO so I can take Virgin. There are so many surprisingly cool things: on-demand food and drinks, comfy leather seats, and good content on the personal screens."

Auby continues: "What keeps me loyal is the staff. Virgin has the coolest flight staff. They never fail to make a 5+-hour flight entertaining. I can tell they love their job. Nothing is worse than grumpy flight attendants."

Steve Naventi, at The Outcast Agency, which represents several high-tech firms like Salesforce.com and Facebook, says:

> Virgin America makes flying feel enjoyable and special for me—like it once did—but with a modern approach. I actually look forward to going to the airport when I'm flying Virgin. The only complaint I've ever had is when I find that their flights won't match my schedule or they don't fly where I'm going!

Billick summarizes:

> Our brand pillars are:
>
> —Advocacy—think of the guest first when acting.

—Innovation—do it better than the other guys, in whatever we are executing.

—Vibe—do it with personality and wit.

He might as well add "continually improve." Virgin has tested free use of the new Google Chromebooks on selected flights.[25] And that also comes with free wi-fi!

Now that is technology-enabled elegance!

Chapter 7

Exponential: Leveraging Ecosystems

Benjamin Pollock was a famous printer of toy theatres in London at the turn of the twentieth century. You can now buy "Pollock's Toy Theatre" as an application on various Apple platforms. It is developed by a company called Zuztertu. Not only is the genre very different from the countless games and other applications in the Apple App Store, but its founder, Gerlinde Gniewosz, has a very different background compared to the typical developer.

Gniewosz has experience at Yahoo! (the web company), Orange (the telecommunications company), and McKinsey & Co. (the strategy firm). She has an MBA from Harvard Business School. She was born and educated in Australia, went to business school in the United States, has worked in Germany and the United Kingdom, and has traveled the world. She's what you might call cerebral—like the portfolio of over 70 Zuztertu mobile apps, aimed at a wide set of markets from apps for small children to education for adults.

At the other end of the spectrum are the applications featured in the GEICO (the insurance company) commercial, which asks, "Do people use smartphones to do dumb things?" It features BroStache, musical instrument, and popping champagne bottle mobile apps as office colleagues at their juvenile best start planning for the weekend and turn their chairs and tables into a rock concert stage.

The BroStache app advertises, "You can choose a 'stache for any occasion—meeting, birthday party, or hot date. Just open the BroStache app, choose a 'stache, then hold it up to your mouth and start talking."[1] Yes, maybe even say something cerebral like Gniewosz would!

The more than 600,000 (as of November 2011) and growing applications in the Apple App Store range from the sublime to the silly. But the sheer size of the marketplace is a formidable Apple asset. The head of iOS development, Scott Forstall, has commented that Apple has, in the iOS App Store, "created the best economy in software in the history of the planet."[2] It is important enough for Apple to go to court to say Apple licenses cover iOS developers, when a company called Lodsys sued some of the developers for alleged patent violation.[3]

Ecosystems—Old and New

From IBM archives: "In 1988 IBM introduced the Application System/400 (AS/400) . . . . As part of the introduction, IBM and IBM Business Partners worldwide announced more than 1,000 software packages in the biggest simultaneous applications announcement in computer history."[4]

The AS/400 went on to be one of IBM's most successful products. Look closely, though. Just 1,000 applications qualified for the biggest announcement in computer history? Compare that to today's slogan for the Apple App Store: "There's an app for that. Hundreds of thousands, actually."[5]

Successful platforms have always boasted large application catalogs, but the scale of the catalog and the downloads in the billions in the Apple App Store and the Android ecosystem are unprecedented.

The Apple/Android Footrace

At Google's I/O conference in May 2011, Hugo Barra, Product Management Director of Google Android, announced to the cheers of 5,000 developers in the audience that Google had exceeded 100 million activations in 112 countries. That ecosystem had over 450,000 developers who had developed 200,000 applications, and users had downloaded 4.5 billion copies. Not bad for a platform that did not exist three years prior! These, by the way, did not include numbers from Android marketplaces that Amazon, Verizon, and Vodafone had started on their own.

The next month, Apple, at its own developer conference, presented even more impressive numbers. Two hundred million iOS devices sold, 425,000 applications, and a whopping 14 billion downloads. And that did not include 15 billion songs sold via iTunes and 130 million e-books sold through iBooks.

Apple and Google both have fans and critics. Gniewosz of Zuztertu, introduced earlier, says: "Personally, I prefer Apple's approach to the app marketplace than Google's approach. When I worked in the telecom industry, I saw firsthand how a lack of quality control on the market offerings led to WAP/Ringtone market revenues dropping to a third of what they were at peak because of scandal and a loss of trust by consumers. The Android platform has a similar lack of quality control at the moment and it will only take a few bad-egg developers to destroy the market for everyone."

In contrast, *BusinessWeek* wrote about Apple: "To keep its developers engaged and loyal, Apple sweats the small stuff. Developers rave about the quality of the company's 'development environment'—the collection of technical specifications, manuals, and programming tools used to write apps."

The article, however, cautioned, "Smaller players wait in frustration to see if Apple's app-approval staff will give their blessing, with almost no means of influencing the decision."[6]

Charlie Wood, whom we introduced in Chapter 5, agrees about the Apple "black hole," its lack of transparency, and is more in Google's camp. "I like the personal relationships with and the support from Google's marketplace team. There are no partner contracts to sign, no red tape, anyone can get started easily. The approval process is painless and relatively transparent."

In the meantime, the war of words between the vendors themselves escalates. At Google I/O, a projected image of Android's green robot chomping into an apple was greeted with thunderous applause. Apple, in turn, mocks the multiple Android stores and says it will only confuse users and developers. In Walter Isaacson's biography of Steve Jobs, Jobs left little doubt he considered Android a "stolen product."

But love them or hate them, both Apple and Google have a blistering lead as Amazon and others try to catch up.

The Amazon, Microsoft, and RIM Catch Up

Amazon opened its own version of the Android app store in March 2011 with 3,800 apps. To get attention, it started with a big promotion, offering Rovio's popular Angry Birds game for free. It needed to start off with a bang to have a shot against Apple, which has more than 50,000 games in the App Store.

Like Apple and unlike Google, Amazon will be reviewing apps before putting them on its store. Unlike Apple and Google, Amazon will be setting the price for the apps, though developers can suggest a retail price.

Amazon, of course, is playing catch-up and has to react to the price leadership of Apple and Google. Wood chuckles about Apple, "The greatest trick the devil ever pulled was convincing developers to sell their apps for 99 cents."

Apple welcomed the Amazon store with a lawsuit for using the name App Store.[7] While that may sound frivolous, it shows the high stakes associated with these ecosystems.

Microsoft "realizes it can't catch up in the near future, if ever. So the company is trying to fuel creation of must-have apps by giving developers free expertise, coding help, phones, and even waiving—for up to one year—the 30 percent sales commission on ads placed on the app that it typically takes."[8]

RIM, as described in the case study at the end of this chapter, is trying to ramp up with several partners, including Microsoft.

The Love-Hate Relationship with the Developer Community

Amazon's decision to control pricing for apps in its store set off The International Game Developers Association. The professional association for game developers sent its members an open letter accusing Amazon of being developer-unfriendly.[9]

Some of the wording included:

> While many other retailers, both physical and digital, also exert control over the price of products in their markets, we are not aware of any other retailer having a formal policy of paying a supplier just 20 percent of the supplier's minimum list price without the supplier's permission.

The love-hate relationship between developers and the platforms they develop for is well documented. Fred Wilson of Union Square Ventures, an investor in Twitter, says, "You should expect that the platform you're building on top of to do something that's not in your interest. If you're going to stay on top of that platform forever you might wake up one day to find that the owner of the platform is competing with you."[10] Twitter launched its own photo service after discouraging developers from launching more client applications.

Newsweek wrote, "Facebook has been finding clever, albeit arm-twisting ways to wring revenue out of the software companies like Zynga that live in its ecosystem."[11]

Nintendo President Satoru Iwata has commented, "I fear our business is dividing in a way that threatens the continued employment of those of us who make games."[12]

Jevon Macdonald, an investor and entrepreneur, offers this advice:

> Customer relationship development is the key to building a healthy company. Within most application ecosystems the customer is treated as a commodity and is effectively traded as such by the platform provider. In return for cheap distribution, small vendors are giving up direct customer ownership and relationship management while ecosystem providers continue to own the customer relationship at its most crucial junctions: activation, acquisition, and retention. Other than distribution, ecosystem vendors receive surprisingly little in

return: no guarantees, no platform stability, and no substantial support. The tradeoff between cheap customer acquisition and customer relationship ownership is significant and should not be underestimated.

Impact on the Enterprise

Dennis Howlett, who blogs for ZDNet, describes a conversation with Bill McDermott, co-CEO of SAP, at the company's event Sapphire Now in Orlando in 2011:

> I was not expecting a great deal from the event. Imagine my surprise when I went for a tour of the mobile applications area. Everything from simple expense receipt handling through to dispensing prescriptions, mobile banking, and complex analytics were shown running on all of the popular devices. iPad, iPhone, BlackBerry PlayBook along with Android devices of every description were shown running mobile applications.
>
> SAP? Who would have thought? It sparked my imagination into wondering what it would mean if SAP opened up its Sybase Unwired Platform and turned it into a developer's playground. I went away and scribbled some "back-of-napkin" calculations in an effort to try to understand what this might mean.
>
> The following day I met with McDermott. I explained my thinking and suggested to him that SAP has in its hands a golden opportunity to capitalize on the potential value of its huge developer network. "Imagine what would happen if SAP had a kind of App store where there were hundreds, thousands of applications. Imagine what it would mean for the democratization of development." McDermott sat there pondering those thoughts and then I showed him my numbers along with comparisons of selling the platform the way SAP usually likes to do things. You know software licenses for a few million dollars plus 22 percent maintenance.
>
> "Which numbers do you prefer?" I asked him. "I like that one over there," said McDermott pointing to a huge figure and beaming from ear to ear. "I can be 90 percent out on any of my calculations and still beat anything SAP will likely get selling the platform to its top 400 customers. So what do you think, Bill?" I inquired. For the first time in all the meetings I've had with McDermott he sat still and said

nothing for a few moments: "I'll need to take that under advisement," he answered, smiling wryly.

Will SAP move away from its traditional partners like Deloitte Consulting and go with individual developers as Apple and Google have? Not just Amazon and RIM, the enterprise vendors and enterprise users have to factor the new ecosystems in their plans.

Impact on Labor Markets

An even more dramatic impact is occurring in the talent marketplace. When Apple shares 70 percent of revenues with a developer, when Google shares 95 percent of Chrome revenues, and they absorb significant marketing, credit card processing, and other fees, it is compelling to become an entrepreneur rather than go work for IBM or Accenture implementing SAP or other enterprise mobile apps.

Ryan Nichols, who used to lead Solution Marketing for Appirio, a mid-sized systems integrator focused on cloud, social, and mobile applications, sees an interesting convergence:

> Today's mobile apps will absolutely inspire a new wave of enterprise productivity, but those productivity gains are likely to be actually delivered by a different set of vendors than we see in the existing (consumer) ecosystem. Developing mobile apps for the enterprise still requires a fundamentally different skillset (and development culture) than developing consumer apps. We'll see a lot of unsuccessful copy/pasting by the incumbent enterprise vendors (SAP is exhibit A here: right ideas, wrong technology). And once again, we'll see a new ecosystem emerge focused on bringing mobile innovation to the enterprise. It's still early days, but Appirio hopes to be a core part of this ecosystem. We'll be inspired by existing consumer app developers, hire from them, and maybe even partner with them, but are convinced that we bring something unique to the table.

App Stores Galore

In Chapter 16, we see the Smart Apps library downloadable on the Lexmark Genesis printer. Garmin has its own store where you can download a variety of maps. Verizon and Vodafone have subsets of the Google Android marketplace and offer subtle differences like allowing customers to add purchases to their monthly billing statement. Nowadays, any smart product has to think about attracting developer communities around them to make them vibrant.

Even the U.S. government has jumped in with its own mobile app store with information about product recalls, jobs, and agencies like the Transportation Security Agency (TSA).[13]

In the next case study we see the approach RIM has taken and adjustments they have made to try to keep up with Apple and Google.

Conclusion

A thick application catalog has always been important for a technology platform's success. It's become dramatically more important in the past few years as Apple and Google, in particular, have shown. In the next section, we will see how RIM has had to react to this new world.

Case Study: RIM's Evolving Ecosystem

"CrackBerry."

That's what BlackBerry users fondly call the devices to which they and their thumbs are addicted. For a while, a BlackBerry was a standard corporate issue in many enterprises, along with an ID badge. President Obama insisted on using one even after entering the White House to the consternation of the Secret Service. RIM, the Waterloo, Ontario-based company that makes the device, was a local darling and a showpiece for Canadian technology.

A few years ago a trend started where people would carry their corporate-issue BlackBerry in one pocket and an iPhone or another smartphone in the other.

And even the standard-issue BlackBerry started to be questioned in many companies. Goldman Sachs analysts explained it as "We believe this is largely due to CIOs following the individual consumer preferences of their employees, similar to the dynamic that has unfolded in the smartphone market."[14]

It reflected the reality that BlackBerries had not evolved in features. Co-CEO Mike Lazaridis is said to have commented, "There will never be a BlackBerry with an MP3 player or camera."[15]

And RIM was just as slow in developing a robust applications ecosystem like Apple had. As corporate interest stalled, consumer interest grew, particularly in the teenage demographic, which likes the free BlackBerry Messenger (BBM) as a substitute for text messaging. (Indeed, in the London riots of summer 2011, BBM earned the dubious distinction as having been the collaboration platform of many of the rioting youth.[16])

As RIM continues targeting the consumer market with smartphones with cameras and other features, and now tablets like the PlayBook, the need for an apps ecosystem has grown even more.

During his opening session keynote at Mobile World Congress in early 2010, Lazaridis introduced the concept of a "Super App":

> Super Apps are the kinds of apps that people love, that they use every day because they offer such a seamless, integrated, contextualized,

> and efficient experience. Our developers are able to build in that enhanced experience because of the unique capabilities of the BlackBerry platform.[17]

The RIM thinking was to let Apple and Google brag about hundreds of thousands of applications; they would focus on the really useful ones. In February 2011, RIM announced there were 20,000 BlackBerry apps available—quite an improvement from a few years earlier but still only a fraction of the volume in the Apple and Android marketplaces.[18]

"Subscribers are facing—on most platforms—thousands of applications to choose from and what's scarce is attention right now. . . . It used to be that services were scarce; now the services are abundant and attention is scarce," Alan Brenner, RIM's SVP of the BlackBerry platform, rationalized in an interview.[19]

He makes a valid point. A Nielsen survey showed the 10 most-used Android apps accounted for 39 percent of all time spent on apps. The total jumped to 61 percent when you account for the 50 most-used apps. That means that hundreds of thousands of apps are competing for the remaining 39 percent of the pie.[20]

Same with the Apple ecosystem. Indeed, Trip Hawkins, the founder of the gaming giant Electronics, calculated that each app in Apple's store earns, on average, about $4,000. He told a conference:

"Four thousand per application: Do you see a problem with that? . . . That doesn't even pay for a really good foosball table."[21]

That may be so, but empty malls are a turnoff, as we all know. As Robert Dutt wrote, "Consumers don't care about 'true multitasking,' 'type to share,' 'synergy,' or surfing with Flash. Businesses care even less about these things. What they care about is what they can do with the device. And today, that's defined more so by the ecosystem of developers around the device than it is by the manufacturers of the device itself."[22]

So, BlackBerry World in Orlando in May 2011 saw quite a turnaround. It started with Adobe CEO Shantanu Narayen announcing that most of the 3,000 tablet apps for the RIM tablet, PlayBook, were built on Adobe AIR.

Adobe, of course, could not wait to respond to Apple's very public banning of its Flash products on its products and Steve Jobs' lashing, "The avalanche of media outlets offering their content for Apple's mobile devices demonstrates that Flash is no longer necessary to watch video or consume any kind of web content."[23]

Then onstage came Microsoft CEO Steve Ballmer announcing that Bing would become the official search and map applications for BlackBerry. Of course, just a couple of months prior, Microsoft and Nokia had announced "the third ecosystem" (the other two being Apple and Android centric) with Nokia deemphasizing its proprietary Symbian and licensing Windows Phone for future products. Microsoft had every incentive to line up even more partners like RIM to regain its once dominant position around mobile phones.

Ballmer was followed by Facebook's vice president of global marketing solutions Carolyn Everson, who announced that the PlayBook would be the first to get Facebook's tablet app.

Adobe, Microsoft, and Facebook—that's an impressive trio RIM lined up. But wait, there's more.

RIM execs also demoed the "Android Player," which enabled "PlayBooks to run apps created for Google's Android OS. Developers can contention their existing Android binaries to BlackBerry App World, and PlayBook owners will download the apps from App World."[24]

Now there are at least four Android marketplaces—Google's, Amazon's, Verizon's, and Vodaphone's. So potentially those Android applications could also be "binaried" over to BlackBerry.

Has RIM overdone it?

Jobs of Apple had already taken a jab at the proliferation of Android stores during a conference call with analysts:

> In addition to Google's own app marketplace, Amazon, Verizon, and Vodafone have all announced that they are creating their own app stores for Android. So there will be at least four app stores on Android, which customers must search among to find the app they want, and developers will need to work with to distribute their apps and get paid. This is gonna be a mess for both users and developers.[25]

There's Android, Adobe, Microsoft, and Facebook in addition to BlackBerry's own ecosystem of applications. So RIM may ramp up its application ecosystem portfolio in a hurry.

How do you effectively set up governance across so many parties? How do you share revenues? Where do you encourage developers to focus? All thorny questions.

In November 2011, Adobe announced it was deemphasizing Flash for mobile devices and moving its focus to the industry standard HTML5. Larry Dignan, Editor-in-Chief at ZDNet, when asked in an interview what this means to RIM, commented:

> Flash support was actually the differentiation point in the beginning for RIM's PlayBook tablet. Adobe's decision changes all of that in a hurry. In the broader context, RIM is fumbling around for an app strategy and any differentiation point. Ultimately, RIM is going to need its own innovation—not just support for technology others control—to regain its luster.

The ecosystem journey RIM has taken is a fascinating one and as it unfolds will determine whether we continue to use the moniker CrackBerry to reflect a central role in our digital lives.

Chapter 8

Efficient: Amidst Massive Technology Waste

"It was the meeting where our IT council gave me the green light to proceed with a major global project to consolidate 80 different HR systems across the world. We had settled on our existing ERP vendor's software," says David Smoley, CIO of Flextronics, one of the world's largest electronics contract manufacturers.

"I thanked them for approving such a large project and then I surprised them. I told them I may come back to them in the next month with a 'better, faster, cheaper' option.

"Mind you, this was after we had already spent months of evaluating options, systems integrators, building budgets and business cases. But we have a 'fail faster' culture here at Flextronics, so they were more intrigued with what I may come back with than surprised."

The Courage to Be More Efficient

In the month prior to the meeting described above, Smoley had been introduced to Workday, then a startup. He says:

> I was immediately impressed by the user interface. It was highly intuitive, attractive and easy to use. Very visual, like Amazon or Google. At the time, usability was one of the biggest issues with the legacy HR packages. I met with the Workday team and got a demo. We discussed technology and I was again impressed by their approach. From my time working in venture capital I knew that some of the earliest and most successful SAAS companies were in the HR space.
>
> After the IT council meeting, the due diligence around Workday grew in intensity. There was strong interest, along with strong resistance. Initially, the head of HR was intrigued because he received positive feedback from his team on the usability factor; however, he was concerned with the risk associated with such a small and early-stage company. The CFO also was concerned with their size and stage. The more I got to know the Workday team, the more I was convinced that while they were early and small, they had the vision, leadership, talent, and experience to deliver in this space. From my perspective, HR IT should be simple. It is basically a database of employees with workflow and reporting associated with it. The legacy HR packages were far too complicated and difficult to use. Workday was an opportunity to start fresh and take advantage of leading-edge technology along with a decade of HR IT lessons the Workday founders had learned at their previous company, Peoplesoft.
>
> We debated back and forth on this and in the end, the tipping point was a Saturday morning meeting at Flextronics, where Aneel Bhusri and Dave Duffield, the co-founders of Workday, met with me and our CEO Mike McNamara. After about 20 minutes we walked out and Mike looked at me and said, 'Those are the kind of guys I want to be in business with.' Duffield and Bhusri had committed to a partnership where Flextronics would provide a benchmark for process and priorities. Flex would drive the development road map for Workday. In turn, Flextronics would draw upon the years of HR experience the Workday team had to shape and standardize its processes. It was a model software development partnership.

> The proposal from Workday saved us more than 40 percent over the life of the project, with even larger savings in the first year. By going with Workday we were able to reduce our team size by more than 60 percent.

"Empty Calories" in Technology

What is remarkable is that Smoley had the courage to turn down a preapproved project and go with something far more efficient. Unfortunately, too much in technology gets funded year after year with "safe" choices. It becomes "entitlement spend." The old adage used to be "No one got fired for buying from IBM." Today, that saying has become a little broader—to include IBM, Verizon, SAP, Accenture, and other large vendors. Too many IT executives live in fear of one of their smaller vendors going out of business.

So, we live in a world of ossified and outrageous technology costs. There are opportunities at every turn to deliver efficiencies in technology, if we only have the courage like Smoley to look at options.

Their estimates vary, but most analysts agree that annual IT and telecom spend exceeds $3 trillion. That's more than the GDP of many nations, and shockingly much of that is "empty calories."

Here are a handful of examples:

> *Software*: Every year, software vendors send customers a bill for 15 to 25 percent of the "licensed value" of their software. This payment is supposed to cover support (bug fixes, help desk) and maintenance (periodic regulatory updates, enhancements). During implementation, few customers tax the software vendor's support lines. In fact, most customers additionally pay a systems integrator or the software vendor's consultants to provide on-site implementation help. Yet they are charged the full software support and maintenance fee. Typically, after a year of going "live" on the software, support needs drop off again. Yet companies continue to be charged full support and maintenance fees. Because the support fees are on the licensed value, users are also charged on the "shelfware"—software licenses the company originally bought but has not deployed. In the

meantime, software vendors have been automating much of the support into knowledge bases so customers can self-service. They are moving support for stable, older releases to their offices in low-cost locations (or using offshore firms to support them). They are increasingly letting user communities handle routine queries. Yet none of these lowered costs have typically been passed along to customers.

Technology infrastructure: The average corporate data center is grossly inefficient compared to the next-generation centers being built by Google, Amazon, and Apple, and certainly compared to the Facebook center discussed in the case study later in this chapter. It is not uncommon to see power usage effectiveness (PUE)—the ratio of power entering a data center divided by the power used to run the computer infrastructure within it—at 2.0, when the Facebook Prineville one is at 1.07. So why can enterprises not walk away from their internal boat anchors? Many are locked into multiyear outsourcing contracts and face stiff early-termination penalties. Meanwhile, there is little in the legal language to force outsourcers to move to more efficient data centers. Walk into an Office Depot or a Curry's and anyone can buy a terabyte of storage for less than $100 and it's a one-time payment. Yet many enterprises are paying $100 or more per gigabyte over a three-year useful life, when you amortize the cost of storage and support for it. Granted, that is high-availability, enterprise-grade storage, but is it worth 1,000 times as much?

Technology services: As companies implement technology projects, they tend to hire systems integrators like Deloitte Consulting. Even though many consultant proposals proudly state they have done hundreds of SAP or Java development projects, typically they do not pass along much in the way of productivity or automation gains. In fact, since they are more experienced, they expect to be paid a premium. But even after companies pay premiums for specialized talent, IT projects still fail at unacceptably high rates. Time-to-completion metrics coming out of newer agile development methods often show a two to three times improvement opportunity compared to traditional delivery models. Another major expense is consultant travel. That often adds another 15 to 25 percent to base fees, which are already high to begin with. The traveling consultants often have a

Monday–Thursday on-site policy, which forces the entire project to adjust to a four-day workweek. Many consultants are implementing telepresence for internal communications but unlike Cognizant, described further on, most have not shown much initiative in using it on client projects and cutting back on project travel.

Telecommunications: Most companies pay a bewildering range of landline, conference calling, calling card, employee wi-fi, mobile, and other messaging fees. Phone companies are also creative with their fees, with shortfall charges, early-termination, and other charges. In most companies, the spend on telecommunications typically exceeds the cost of all other technology costs—hardware, software, technology services, internal staff—put together. Not surprisingly, there is plenty of waste. One of the most flagrant is in international mobile calling charges. A study by Harris Interactive showed that the average U.S. employee spends $693 in international roaming calls on an overseas trip at between $1 and $4 a minute when using local carriers; Skype would cost just pennies.

Other technology costs: Many printer inks cost over $5,000 a gallon (yes, over 1,000 times the price of gasoline). Ink is far more expensive, ounce for ounce, than the most expensive perfumes or rare spirits. Worse, the ink gauges can be inaccurate and force consumers to discard cartridges before they are empty. Another major cost is technology strategy and market intelligence. Peer benchmarking opportunities and market intelligence via boutique research firms like Redmonk (whose analysts are quoted in several chapters in this book) and decent quality but free blogs, such as ZDNet, GigaOmM, TechCrunch, and other sites quoted in several chapters, are helping companies lower costs of IT market intelligence.

There is a screaming need for efficiency in technology, and it is gratifying to see enterprises that run their technology in a lean fashion.

Efficiency in Government

Nestled between a Target and a Lowe's in a Tampa, Florida, shopping center is a much smaller nondescript office. During some hours, this office does more business than either of those two giant retailers. It is the Hillsborough County Tax Collector's branch office. It issues driver's licenses, collects property taxes, issues fishing licenses, and does several more things. While there are separate queues for many of the services, the agency increasingly has employees who can handle multiple transactions in one session. Of course, for many of the transactions, you need not come in. You can do business with them by mail, phone, or on the web.

Doug Belden's site says, "His goal as tax collector is simple: To save taxpayers money through consolidation and efficiency while improving service at the branch offices. His objective is to make the Hillsborough County Tax Collector's Office the most modern and efficient office in the state."[1]

Walk into that branch, and there usually is a mass of humanity, and your ticket number may be in the 600s. Your heart sinks as you get ready to settle in for a few hours.

Twenty minutes later your number comes up. Five minutes after that the agent has scanned four pieces of personal documentation you have brought. In another five minutes you have a polycarbonate driver's license with digitized photo, state holograph, two-dimensional bar code, and magnetic strip. In those five minutes, the system has been running validations against various state and federal databases. These are part of the checks under the Federal Real ID legislation requirements, and for that, licenses issued by the county are among the first in the country to carry a gold star, which shows compliance.

The customer queuing system is from Q-Matic. Beyond balancing customer load with agent availability, it is smart enough to route Spanish-speaking customers to appropriate agents. It collects all kinds of metrics, which are then helpfully summarized, and before you leave home, you can check likely wait times at each of the agency's locations.

Beyond what is visible to the ordinary, Kirk Sexton of the agency is happy to talk about other efficiencies. There are VoIP and virtualization efforts. At the back there is a microwave tower that makes communications cheaper than a T1 line. Sexton talks about a penny-a-page arrangement with SunPrint Management (described further in Chapter 11) for their many black-and-white copies. He describes buying mostly standard, off-the-shelf equipment to get volume breaks—but making adjustments as needed. The Samsung printers had their firmware adjusted to control their speed to optimally heat the car license tag.

No, this is not your run-of-the-mill local agency. In 2008 it won the Florida governor's prestigious Sterling Award. In 2010 it applied to be considered for the national Malcolm Baldrige Award. When you look at the applications the agency filed for consideration for both awards, you see the kinds of eye-popping metrics even private sector firms would kill for, focused on customer satisfaction, customer wait times, and economics.

The Tax Collector's office is an elected one, but Belden has not had much serious competition for 12 years. As we can see, he runs a tight ship with a continuous improvement focus. With folks like Sexton who joined the agency three years ago from the private sector, it is clearly a role model for efficiency and innovation in government.

Efficiencies Even When Things Are Going Well

Cognizant is a leading IT, consulting, and business-process outsourcing services provider. It was spun out of Dun & Bradstreet in the mid-1990s and has grown like a weed ever since. As its website succinctly summarizes, "A member of the NASDAQ100, the S&P 500, the Forbes Global 2000, and the Fortune 500, Cognizant is ranked among the top-performing and fastest growing companies in the world."[2]

Given its phenomenal success, you would think it would not be as focused on efficiency. Two of its executives, in particular, highlight a relentless focus on productivity and efficiency.

Gordon Coburn, its CFO (and COO and Treasurer), has plenty to keep him busy between analyzing and integrating acquisitions and handholding investors. On any given day, though, you will see him intensely involved in real estate negotiations, from square footage cost through space design and related accoutrements (like wall and carpet colors), and in software and other technology negotiations. He is a bulldog in such negotiations.

Sukumar Rajgopal wears two hats. He is the Chief Information Officer and worries about infrastructure for hundreds of offices and over 100,000 worldwide employees. He is also the Chief Innovation Officer and is constantly thinking about how to make the company's consultants more productive—*x* times more efficient.

Cognizant's global footprint means hundreds of meetings every week among associates working from widely dispersed locations. Project teams are increasingly virtualized, with many individuals, associates, and client staff working on closely related tasks from different sites. So its investment in telepresence has been a major boon allowing Cognizant teams to collaborate.

Cognizant had originally planned to equip 20 major offices with full-immersion, three-screen systems for internal meetings. They quickly learned that the full-immersion terminals were less frequently used than a single-screen setup in a conference room or on a desktop. So they shifted gears to scale telepresence swiftly across their global footprint with the smaller devices, many from Tandberg (now part of Cisco).

The initial deployment was equipped on desktops of the senior management team, especially in home offices. Coburn remarks that "telepresence has become the standard for how we do senior management calls." It also has become the preferred approach for early-stage interviews with recruits. Cognizant has also started using telepresence for most meetings in its budgeting process. As Coburn explains, "It made a tremendous difference; it probably shortened the budget cycle by 25 percent, because there was far less confusion."

More impressively, Cognizant has rolled telepresence out to major clients. That helps overcome at least some of the consultant travel costs and other issues mentioned earlier. Clients were generally skeptical at first, especially if they had previously used traditional videoconferencing, which was characterized by high cost, frequently dropped calls, and video and audio quality that ranged from poor to mediocre. But they have been much more accepting of telepresence, which delivers a vastly superior user experience.

Two areas with high payback have been during transition and training. *Transition* is the transfer of a client's operation (e.g., software maintenance or call center support) to Cognizant people and infrastructure. In the past, this typically required Cognizant specialists to be at a client site for several weeks. The start date was often determined by travel logistics, visa availability, and so on. But with telepresence, they can start the transition with less delay, involve more specialists, and do it at lower cost than before.

When Cognizant brings a client's process in-house, it often needs to conduct extensive training. This typically requires a trainer to travel, with all the same visa and logistics issues. With telepresence, it can conduct the training with less delay and at a lower cost than before.

It shows in a significant reduction in Cognizant's emissions associated with business travel. They decreased from 35,964 metric tons of CO2e (Carbon Dioxide equivalent) in 2008, to 27,738 in 2009, or a 23 percent reduction, even as business grew by 16 percent over that time period. Although some of this was due to controls of travel costs, telepresence was instrumental in driving these significant results.

Sukumar, on the other hand, is obsessed with what he calls "social design," and his quest to deliver 500 percent productivity on Cognizant projects.

> Think of avalanches. A snowball starts small, but then gathers mass and gradually turns into a massive avalanche. That is the basic inspiration behind Social Design. What we do individually should (positively) impact hundreds and thousands of others.
>
> One of my favorite examples is CDDB (short for Compact Disc Database). When an individual ripped tunes from a CD, before CDDB you had to manually track names, artist, etc. Each user around the world did that and did in their own format with their own typos. CDDB started to track unique signatures of each tune on its servers and matching them to album, artist, and other information. So when later users ripped the same tune, they could download that same information. Think of the massive productivity that delivered across millions of users.
>
> Today's enterprise systems don't acknowledge like CDDB did that we live in a connected world and what one user enters can be used to populate that for thousands of others. Our systems record, they don't think. When you enter your time sheet, the system should prefill a number of fields and it should give you feedback like "you are the last one in your office to file it" or "you forgot to enter a few items others in your group may have entered."
>
> It would be nice to move to a "productivity income statement" where we charge the IT department $1 for each data item users enter and in return we charge users $1 for each data item that the system

prefills for them. Think how differently we would think about enterprise systems if we did that.

As we discussed in Chapter 2, Apple, in announcing its iCloud Match service, provided another example of what Sukumar calls social design.

"We have 18 million songs in the music store. Our software will scan what you have, the stuff you've ripped, and figure out if there's a match."[3]

Apple's site helpfully added, "And all the music iTunes Matches plays back at 256-Kbps iTunes Plus quality—even if your original copy was of lower quality."[4]

The CDDB and Apple examples are both examples of personal productivity. Imagine if Sukumar can inspire Cognizant consultants and clients to deliver similar productivity gains in corporate applications.

Conclusion

The technology elite don't just focus on innovation to improve the top line. They are also intensely focused on efficiencies. In this case study we will see the efficiencies Facebook has delivered in its Prineville data center it opened in 2011.

Case Study: Facebook's Hyperefficient Data Center

"Our server chassis is beautiful," says Amir Michael, Manager, System Engineering at Facebook.

That's beautiful in a minimalist kind of way. "It's vanity-free—no plastic bezels, paint, or even mounting screws. It can be assembled by hand in less than eight minutes and thirty seconds."

Michael is describing the design principles behind Facebook's data center, which opened in April 2011 in Prineville, Oregon (previous data centers were leased). The center occupies 150,000 square feet to start, with another 150,000 square feet in progress. It handles about half the processing of Facebook's staggering user base demands—over 700 billion minutes a month are spent on Facebook.[5] Facebook says it has 750 million active users (as of July 2011), 70 percent of which are outside the United States and access the site in over 70 languages.[6]

The servers, Michael describes, are almost six pounds lighter than the models Dell or HP (Facebook suppliers for the first phase of the project) sell to other customers. That, of course, lowers shipping costs and eases technician effort to move them around. The lack of covers also allows easier technician access to components and more direct cooling. That all adds up when you are talking thousands of servers. Facebook says in a typical data center this would save more than 120 tons of materials from being manufactured, transported, and ultimately discarded.

Efficiencies Galore

Everything in the center's design was driven from a cost and efficiency perspective, and most of the components were designed from the ground up to Facebook's specifications.

The chassis, for example, is 1.5U form/factor (2.63 inches tall) compared with standard 1U (1.75 inches tall) chassis. That allows for larger heat sinks and larger 60 mm fans as opposed to 40 mm fans.

Larger fans use less energy to move the same amount of air. That takes 2 to 4 percent of the energy of the server, compared to 10 to 20 percent for typical servers. The heat sinks are spread at the back of the motherboard, so none of them will be receiving preheated air from another heat sink, reducing the work required of the fans.

The racks are "triplet" enclosures that house three openings for racks, each housing 30 of the 1.5U Facebook servers. Each enclosure is also equipped with two rack-top switches to support high network port density.[7]

Prineville enjoys dry, cool desert weather at its altitude of 3,200 feet. In what is likely good application of feng shui, the data center is oriented to take advantage of prevailing winds feeding outside air into the building. The data center thus takes advantage of the free, natural cooling, and in winter, heat from the servers can be used to warm office space. On warmer days, when they cannot use the natural cooling, they use evaporative cooling. Air from the outside flows over water-spray-moistened foam pads. There are no chillers on-site—common in most other data centers, thus saving significantly on capital and ongoing energy to run them.

Backup batteries, which keep servers running for up to 90 seconds before backup generators take over, are distributed among server racks. This is more efficient because the batteries share electrical connections with the computers around them, eliminating the dedicated connections and transformers needed for one large store. This design uses only about 7 percent of the power fed into it, compared to around 23 percent for a more conventional, centralized battery store approach.

The motherboards are bare bones and devoid of features Facebook did not need, and geared to handle processors from AMD and Intel. They came with direct connections to the power supply, which itself was uniquely designed.

James Hamilton is considered a "Jedi Master" in data center circles, with experience at IBM, Microsoft, and now at Amazon Web Services. He was invited to visit the Facebook data center as it opened, and he analyzed its features on his blog. Here is what he wrote about the power supply and related backup UPS (Uninterrupted Power Supply) infrastructure:

> In this design the battery rack is the central rack flanked by two triple racks of servers. Like the server racks, the UPS is delivered 480VAC 3-phase directly. At the top of the battery rack, they have control circuitry, circuit breakers, and rectifiers to charge the battery banks. What's somewhat unusual is the output stage of the UPS doesn't include inverters to convert the direct current back to the alternating current required by a standard server PSU. Instead the UPS output is 48V direct current, which is delivered directly to the three racks on either side of the UPS. This has the upside of avoiding the final invert stage, which increases efficiency.[8]

To net that out, Facebook believes this proprietary UPS design (for which it has applied for a patent) will save up to 12 percent in electricity usage.

Adding up all the efficiencies Facebook implemented in the data center, it says it delivered a 38 percent energy efficiency and a 24 percent lower cost compared to comparable existing facilities. The Data Center is rated at a power usage effectiveness (PUE) of 1.07—one of the best in the industry and much better than the 1.5 in its previous facilities. PUE is an indicator of data center energy efficiency—how much of power input goes toward computing versus cooling and other overhead activities—and the lower the number, the more efficient.

The Open Compute Thinking

But Hamilton's next statement makes the center really stand out. "What made this trip (to the Prineville data center) really unusual is that I'm actually able to talk about what I saw."

He continues:

> In fact, more than allowing me to talk about it, Facebook has decided to release most of the technical details surrounding these designs publically. In the past, I've seen some super interesting but top secret facilities and I've seen some public but not particularly advanced data centers. To my knowledge, this is the first time an industry leading design has been documented in detail and released publically.[9]

Facebook calls it the Open Compute Project. They elaborate: "Facebook and our development partners have invested tens of millions of dollars over the past two years to build upon industry specifications to create the most efficient computing infrastructure possible. These advancements are good for Facebook, but we think they could benefit all companies. The Open Compute Project is a user-led forum, to share our designs and collaborate with anyone interested in highly efficient server and data center designs. We think it's time to demystify the biggest capital expense of an online business—the infrastructure."[10]

Facebook is publishing specifications and mechanical designs for Open Compute Project hardware, including motherboards, power supplies, server chassis, and server and battery cabinets. In addition, Facebook is making available its data center electrical and mechanical construction specifications.

This is a remarkable move on the part of Facebook. As we will discuss in Chapter 13, Apple's iCloud data center was not even visible on Google Earth until a few days before the iCloud public announcement. A giant 500,000-square-foot facility was kept "hidden." Google, Microsoft, Amazon, and others have also traditionally been secretive about their operations.

Frank Frankovsky, Facebook's director of hardware design, was quoted as saying, "Facebook is successful because of the great social product, not [because] we can build low-cost infrastructure," he says. "There's no reason we shouldn't help others out with this."

What is interesting is Dell's Data Center Solutions business says it will design and build servers based on the Open Compute Project specification. Presumably so will HP's. Dell owns Perot and HP owns EDS, and those outsourcers currently run data centers for many corporations but they are nowhere as efficient as what Facebook has built. In the meantime, Facebook is planning to source servers even more efficiently by going direct to Original Design Manufacturers who supply HP and Dell.

George Brady, Executive Vice President, Technology Infrastructure, Fidelity Investments, was quoted as saying, "Data centers provide the foundation for the efficient, high quality services our customers have come to expect. Facebook has contributed advanced reference designs for ongoing data center and hardware innovation. We look forward to collaborating with like-minded technology providers and partners as we seek ways to learn from and further advance these designs."

The Green Question

One quibble that environmentalists had with the Facebook center is that it is fueled by the utility Pacific Power, which produces almost 60 percent of its electricity from burning coal. Greenpeace ran a campaign for Facebook to "unfriend" coal. As we discuss in Chapter 17, Google, through its energy subsidiary, has negotiated several long-term wind-power agreements. It sells that energy in the open market at a loss but strips the Renewable Energy Credits and applies them for carbon credit to the conventional power it also uses to run its data centers. Toward the end of 2011, Greenpeace won a moral victory as Facebook promised a preference for access to clean and renewable energy in picking future sites for data centers. Facebook has also recruited Bill Weihl, formerly Google's "Energy Czar."

Barry Schnitt, Facebook's director of policy communications, provides an alternative perspective on clean energy: "As other environmental experts have established, the watts you never use are the cleanest and so our focus is on efficiency. We've invested thousands of people hours and tens of millions of dollars into efficiency technology and, when it is completed, our Oregon facility may be the most efficient data center in the world."[11]

As Facebook's Michael would say, "Now that's a beautiful thing."

Chapter 9

Mobile: If It's Tuesday, It Must Be Xiamen

The National Science Board's Science and Engineering Indicators 2010 report is chock full of facts and figures about the global technology landscape and it summarizes:

> Science and technology (S&T) are no longer the province of developed nations; they have, in a sense, become "democratized." Governments of many countries have firmly built S&T aspects into their development policies as they vie to make their economies more knowledge- and technology-intensive and, thereby, ensure their competitiveness in a globalizing world. These policies include long-term investments in higher education to develop human talent, infrastructure development, support for research and development, attraction of foreign direct investment and technologically advanced multinational firms, and the eventual development of indigenous high-technology capabilities."[1]

In fact, we live in a world where economists are furiously rethinking global trade and deficit definitions to handle scenarios like these:

> By any definition, the iPhone belongs to the high-tech products category, where the U.S. has an indisputable comparative advantage. In effect, the PRC (China) does not domestically produce any products that could compete with iPhones. The U.S. also has an absolute advantage in the smartphone category. Ricardian theory and Hecksher-Olin theory would suggest the U.S. should export iPhones to the PRC, but in fact the PRC exports iPhones to the U.S. All ready-to-use iPhones have been shipped to the U.S. from the PRC. Foreign direct investment, production fragmentation, and production networks have jointly reversed the trade pattern predicted by conventional trade theories. The manufacturing process of iPhones illustrates how the global production network functions, why a developing country such as PRC can export high-tech goods—at least according to the currently applied methodology for calculating trade statistics—and why the U.S., the country that invented the iPhone, becomes an importer."[2]

Of course, if it was just iPhones and iPads, it would be one thing. China has become the technology manufacturing, or at least the assembly, hub for the world. More broadly, BRIC (Brazil, Russia, India, and China) has become a commonly discussed acronym in technology circles.

The "BRIC Wall"

Max Henry graduated from France's ISG business school, but he has spent the past two decades in Asia. He is a publisher of various Asian technology magazines. In a recent issue of CHaINA magazine the focus was on two second-tier Chinese cities—Xiamen and Qingdao.[3]

In many ways, though, Henry is a high-end tour guide to facilities in such cities. Every few weeks his Global Supply Chain Council organizes field trips to plants and ports across China. It could be to Juijang on the southern shores of the Yangtze River with visits to the solar photovoltaic manufacturer Sornid and the helicopter manufacturer Red Eagle. Another trip could be to Qingdao (better known to many in the West as Tsingtao—as in the beer), host to Haier Group, one of the world's largest home appliance manufacturers, and GERB, which focuses on vibration control in heavy machinery, equipment, structures, and trackbeds. Perhaps it could be to Xiamen on the southeastern Chinese coast where Dell has a factory for the booming local China market as does Philips Lighting Electronics, which is focused on next-gen LED and other lighting. Or it could be to Chengdu in western China, home to Intel assembly and testing facilities, and to Giant, the largest bicycle manufacturer in the world.

Western China is poised to become even more influential, as China seeks to build out the next generation of the Orient Express. A few years ago it would have been a pipedream, but with China aggressively deploying high-speed trains within its borders, the possibility of extending them to connect Asia and Europe looks more and more realistic.

As we saw in the HP case study in Chapter 2, Asia and Europe are excited about a rail service between Chongqing and Duisburg, Germany, covering nearly 7,000 miles in 13 days. That is 26 days quicker than the current rail-sea combination, and considerably cheaper than the air option. It should get even quicker as China shares its growing high-speed rail experience with countries along the path.

A research paper looked at the economics compared to the alternatives available today and estimated rail transport could be significantly cheaper than the commonly used combination of air and sea.[4]

China, of course, is going through explosive growth and has its share of inflationary and labor issues. In a survey eight years ago, the American Chamber of Commerce in South China found that 75 percent of its members were focused mainly on export markets. By last year, that number had flipped: 75 percent of 1,800 respondents now say their manufacturing operations in China are focused on serving the Chinese market. That's mainly because China's workers are steadily getting richer.[5]

Foxconn, the largest electronics contract manufacturer in the world, has over a million employees in China.[6] It assembles iPads and other devices for Apple (and has brought Apple scrutiny with a spate of employee suicides and plant explosions). It does similar work for many other high-tech companies. Given labor dynamics in China, Foxconn is exploring a diversification with sizable investment in Brazil.[7] Even more significantly, it is looking to robots to do simple and routine work such as spraying, welding, and assembling. "The company currently has 10,000 robots and the number will be increased to 300,000 next year and 1 million in three years."[8]

China is increasingly also looking to move up-market. "The U.S. has worried about China producing cheap goods—they really should be worried about China producing better goods," said Bruce Rockowitz, the chief executive of Li & Fung, the largest trading company supplying Chinese consumer goods to American retail chains.[9]

Brazil is looking to replicate China's success with high-tech manufacturing and R&D.

"At first glance Florianopolis, in southern Brazil, resembles the quintessential picture postcard resort. It has become one of South America's most popular destinations, a magnet for sun seekers. But for all its hedonism, Florianopolis has its gaze firmly fixed on something altogether more serious—becoming the regional technological powerhouse. Sapiens Parque science park is the $1.3 billion (£81m) brainchild of the Brazilian government, as part of a wider $24 billion (£15bn) initiative to promote science and technology in the country."[10]

Russia is also pitching its wares: "Boeing, Cisco, Intel, Microsoft, and MIT are key partners in Russia's high-tech Skolkovo Innovation Center (Skolkovo), which opened yesterday its American beachhead in Silicon Valley along with OAO ROSNANO and the Russian Venture Company (RVC). The new Russian Innovation Center will serve as U.S. representative office of the three organizations. It will promote and coordinate high-tech partnerships and scientific cooperation in the IT, biomedicine, energy efficiency, nanotech, nuclear, and aerospace sectors among top Russian and American companies, venture capital firms, and academic and scientific institutions."[11]

Then there is Novosibirsk. Already a major Russian high-tech hub, if the northern branch of the China–Europe train becomes reality, its Siberian remoteness could become much less of an issue. "None of our programmers in Novosibirsk are programmers by education," says Intel's Chase. "They are physicists, chemists, biologists, mathematicians. They are first of all scientists. Secondly, they learn how to program, as an afterthought. This combination is extremely powerful."[12]

India is taking a slightly different tack. With its software competency, it is much better positioned for the global R&D/product engineering services market, which Mukesh Dialani, analyst at the research firm IDC, expects to reach an estimated $57 billion by 2014. He defines those services as "the taking over of the research and development of a product company's value chain (in part or full) by a third-party service organization."

Dialani provides examples of such services:

> *Embedded technology services.* This includes writing software called firmware and burning it on a read-only memory (ROM) device, flash memory, or integrated circuit (IC). Embedded systems find applications in the following industries: avionics/aerospace, automobiles, consumer electronics, telecom including mobile and wireless, medical devices, semiconductors, office automation, and industrial electronics.
>
> *Semiconductor services.* These services involve capabilities to provide end-to-end solutions from design through implementation in the area of printed circuit boards (PCBs) and very large-scale integration (VLSI) design. VLSI enables increased functionality in smaller ICs and also customizes the IC, resulting in the creation of application-specific integrated circuits (ASICs) for specific applications, rather than a generic IC that finds application across multiple industries and products.
>
> *Plant design and automation.* Plant design involves the use of multiple engineering faculties, including civil, mechanical, and electrical, to create a design for a process plant. Subsequent to the design phase, service providers will recommend and/or assist with products and services to automate the plant. This involves the use of many technologies that could include robotics, control systems, instrumentation systems, sensing equipment, motors, and valves.
>
> *Software product development.* The service provider will gather the requirements from the customer and create a specialized or niche product. As an example, the ability to access software over the web could be a very good feature and increase customer adoption.

These are very different from IT services Indian offshore firms have typically specialized in. Many customers started with them for Year

2000 remediation and have used them since for their SAP, Oracle, and other application maintenance. Many customers want a "separation of church and state." They want the product manufacturer/assembler like a Foxconn to be separate from the design team.

India-centric vendors such as HCL Technologies, QuEST, Symphony Services, Wipro, Persistent Systems, and others have sizable product engineering practices. HCL says it has been involved in all major commercial aerospace programs around the world in the past decade, including the Boeing 787 program described in the case study in this chapter. It has also created a smart home automation solution AEGIS, which is a multi-platform gateway that controls, among other things, multimedia, energy monitoring, HVAC, and security. QuEST provides companies like Caterpillar 3D modeling and other product engineering services. Symphony Services does product development work for a wide range of technology vendors like Adaptec, Lawson Software, and Motorola. Wipro has a joint venture with GE Healthcare and a large set of patents around wireless networks and Bluetooth technologies. Persistent Systems, which develops software for over 300 customers, most of them technology vendors, says it has delivered over 3,000 software product releases in the past five years.

Beyond BRIC

Over the past few years, Accelerance, based in Silicon Valley, has seen U.S. demand for offshore software development shift markedly from traditional markets such as India to nearshore destinations like Costa Rica, Colombia, Mexico, and Argentina. "For small to medium-sized U.S. companies, nearshore is most often the way to go," says Andy Hilliard, President of Accelerance. "Nearshore software development firms have had time to benchmark best practices from offshore and they provide attributes like proximity, time zone alignment, and cultural compatibility that offshore can't match. Though nearshore rates are 10 to 20 percent higher than offshore, on a project basis clients can still save in excess of 50 percent, while significantly increasing their control and ability to collaborate on projects." Accelerance has developed a Global Software Outsourcing Network that provides clients exclusive access to prescreened offshore or nearshore outsourcing providers in over two dozen countries. As an example, Accelerance helped the software vendor Ariba partner with a firm in Costa Rica. Ariba had plenty of experience developing software in the United States and India, but it wanted to distribute its agile software development efforts and diversify its global software development locations beyond India.

EPAM Systems offers several software customers like Oracle, Microsoft, Novell, and SAP an Eastern European option with a wide range of talent in Russia, Hungary, Poland, Belarus, and Ukraine.

The Japanese tsunami in early 2011 showed how many components in high-tech devices still come from Japan. "The IHS iSuppli teardown analysis of the iPad 2 so far has been able to identify five parts sourced from Japanese suppliers: NAND flash from Toshiba Corp., dynamic random access memory (DRAM) made by Elpida Memory Inc., an electronic compass from AKM Semiconductor, the touch screen overlay glass likely from Asahi Glass Co., and the system battery from Apple Japan Inc."[13]

Looking ahead, Apple set off all kinds of market speculation when it announced in early 2011 it would spend about $3.9 billion as prepayment for components from three companies, without specifying the suppliers or the parts. The research firm iSuppli speculated it was for displays made by LG, Toshiba's screen unit, and Sharp. These are Korean and Japanese companies.

Foxconn, mentioned earlier, is Taiwan-based, and a number of other Apple components come from Taiwanese suppliers. Flextronics, another large contract manufacturer like Foxconn, has Singapore roots. Jabil, another contract manufacturer headquartered in Florida, has shown adeptness in using facilities in diverse locations such as Mexico and Malaysia. Both Flextronics and Jabil also provide a wide range of warranty and repair services for their customer products. Then there are other Taiwanese players like Pegatron, a spinoff from Asus and Quanta.

The National Science report referenced previously uses the term Asia-9 as promising science and technology countries—India, Indonesia, Malaysia, the Philippines, Singapore, South Korea, Thailand, Taiwan, and Vietnam. The report also compliments Israel, Canada, Switzerland, and South Africa.

Behind the phenomenal success of the Microsoft Kinect is an Israeli company, PrimeSense, which licenses the hardware design and chip used in the Microsoft product.

The Globalization of Data Centers

For services that Google, Amazon, Apple, and others provide for our searches, streaming music, storage, games, and other applications, we are seeing other countries factor as locations for their data centers.

Google has turned a former paper mill in Hamina on the Gulf of Finland into a data center. "It had to trawl back through 30 years of seawater temperature records, and employ a large amount of thermal modeling to ensure it could cater for the effects of the wind, direction of the tide and ebb and flow, as well as seawater temperatures and density of the seawater before it could devise a solution that would work for a data center. It also had to consider seawater's corrosive properties."[14]

Similarly, Google decided on a data center in Belgium where the climate "will support free cooling almost year-round, according to Google engineers, with temperatures rising above the acceptable range for free cooling about seven days per year on average. The average temperature in Brussels during summer reaches 66 to 71 degrees, while Google maintains its data centers at temperatures above 80 degrees. So what happens if the weather gets hot? On those days, Google says it will turn off equipment as needed in Belgium and shift computing load to other data centers. This approach is made possible by the scope of the company's global network of data centers, which provide the ability to shift an entire data center's workload to other facilities."[15]

Ireland has emerged as a major hub with the growth of cloud computing. It hosts "huge data centers powering the global cloud operations for two of the largest players in the sector, Amazon.com and Microsoft. Dublin also hosts major data centers operated by Digital Realty Trust and facilities for IBM and SunGard, among others."[16]

Dell is building a network of data centers around the world, including one in Australia, as it grows its own cloud services.[17] Japan, Singapore, and Hong Kong contain most of the data centers in the Asia-Pacific region. "While the facilities in Japan are mainly to support its domestic market, Singapore and Hong Kong are competing to be the preferred location for hosting data centers for service providers that serve MNCs in the region."[18]

Dell will find it follows a path already taken by Interxion, which was established in 1998 in the Netherlands. The company manages 28 data centers, covering 13 cities in 11 countries.

What Old World?

Across the world, technology is allowing cities to reinvent themselves. A prominent example is Eindhoven in the Netherlands. It is called the City of Lights because of its much longer association with Philips Lighting than Xiamen described earlier. More recently, it has earned the title of the world's "smartest city" as a protoype of a Western city learning to stay relevant in the world of BRIC and Korea and Taiwan.

> Eindhoven is a manufacturing center in a high-cost country. By focusing on producing high-value, technology-based products, it is in competition with fast-growing manufacturing centers in nations with much lower costs. Many are striving mightily to perfect the complex manufacturing capabilities that have made Eindhoven successful, which creates unceasing pressure for the region to boost productivity. Foreign competitors are also seeking to raise their own game in R&D and knowledge creation, and Eindhoven, which generates 50 percent of all Dutch patents, needs to stay ahead of the curve.
>
> At the same time, however, Eindhoven is saddled with Europe's demographics, in which a low birth rate and aging population is reducing the regional labor force. To win the battle for the talent that provides its competitive advantage, the region must make itself economically and socially attractive to knowledge workers from around the world.
>
> Eindhoven's answer to these challenges is a public-private partnership called Brainport Development (www.brainport.nl). Its members include employers, research institutes, the Chamber of Commerce, the SRE, leading universities, and the governments of the region's three largest cities. A small professional staff meets regularly with stakeholders to identify their strengths, needs, and objectives, then looks for opportunities for them to collaborate on business, social, or cultural

> goals. Any stakeholder of Brainport has the opportunity to create new initiatives or partner with other stakeholders. Their work is based on a strategic plan called Brainport Navigator 2013 (with a 2020 version in the works funded in part by the Dutch government). It calls for focusing on five key areas for development: life technologies, automotive, high-tech systems, design, and food and nutrition.[19]

Beyond Eindhoven, the National Science report above speaks well of the European Union in general: "Its innovation-focused policy initiatives have been supported by the creation of a shared currency and the elimination of internal trade and migration barriers. Much of the EU's high-technology trade is with other EU members. EU research performance is strong and marked by pronounced EU-supported, intra-EU collaboration. The EU is also focused on boosting the quality and international standing of its universities."

Conclusion

To become a tech elite, you have to be a Marco Polo and a Gulliver—bravely explore a fast-changing world. Suppliers and captive units in exotic locations can provide unique competitive advantages. But they can also make much more complicated the supply chain and product development cycles. Let's next look at how Boeing used a global supply chain for its new 787, and how it used HCL Technologies as a "glue" to bond many of those widespread elements.

Case Study: The Boeing 787 and HCL Technologies

You are driving north of Seattle and hear a giant Boeing 747 approaching. Then you do a double take. This one does not have the distinctive hump of a 747. The hump actually extends across most of the length of the plane. It is a Dreamlifter, a specially modified 747 with a massive cargo capacity of 65,000 cubic feet and takeoff weight of up to 800,000 pounds.

The Dreamlifter is the primary means of transporting major portions of the new Boeing 787 from suppliers in Italy, Japan, and many other countries to the final assembly site in Everett, Washington. This has allowed Boeing to reduce delivery times to as little as one day from as many as 30 days it would normally take via sea and ground transport. This is one of countless innovations Boeing implemented around the manufacturing of the 787—branded the Dreamliner.

The involvement of so many suppliers across the globe has been criticized, especially given the repeated delays and related increased costs on the 787 program. The reality is Boeing's main competitor Airbus had already shown that large sections of a plane could be made in four countries—Britain, France, Germany, and Spain—and then assembled in France or Germany.

Boeing CEO Jim McNerney has acknowledged, "In retrospect, our 787 game plan may have been overly ambitious, incorporating too many firsts all at once—in the application of new technologies, in revolutionary design-and-build processes, and in increased global sourcing of engineering and manufacturing content."[20]

When the 787 program was kicked off in early 2003, Boeing was adjusting to new market realities.

> As rising fuel costs and even terror attacks have impacted airlines, causing major swings in aircraft orders, OEMs like Boeing have realized that the up-front investments in tooling required to build a new plane represent growing financial risks. At the same time, experts say, aircraft lifecycles have begun to shorten, increasing the odds that Boeing's

> integrated approach to design and integration could create bottlenecks. So, with the 787, Boeing decided to spread the risk among its suppliers and to call on them to help shorten cycle times.[21]

The 787 plane itself has many innovative features, and the savings expected from a streamlined supply chain allowed Boeing to price aggressively. That made the 787 the fastest selling airliner ever designed. By 2007 Boeing had announced firm orders for 544 airplanes from 44 airlines.[22] While the delays since have led to several canceled orders, Boeing says it is sold out through the end of the decade. In late 2011, as Boeing showcased the plane as part of a launch "Dream Tour," there were enthusiastic crowds across the globe.

The use of composite materials in the fuselage and the wings, the sweptback aerodynamic wings, and the more efficient engines are expected to combine to deliver 20 percent better fuel performance than a similar sized 767 of today. The plane promises flight ranges of up to 8,500 nautical miles, unusual for a mid-sized jet but critical as passengers increasingly demand non-stops across the globe.

And passengers enjoy improvements, such as:

- The LEDs allow the crew to adjust the lighting to match different phases of the flight. The light is fairly standard during boarding and while cruising. During meals it is adjusted to warmer tones. Once you're done eating and want to tilt the seat back and relax, the cabin can be bathed in a relaxing lavender hue. When it's time to sleep, the lights are turned way down.[23]
- The cabin will feel less dry (humidity twice as high as on current planes), because the 787 cooling system will be driven by electricity. An electrical system makes it easier to humidify the cabin air because it's not starting with the hot, dry air from the jet engines common in most planes.[24]
- The superior strength of the composite fuselage will allow the passenger cabin to withstand higher pressurization at an altitude of 6,000 feet instead of the usual 8,000 feet. Passenger comfort is shown to increase significantly at lower cabin altitude pressure.[25]
- An active gust alleviation system will improve ride quality during turbulence.

A major innovation passengers will not see but will benefit from is Boeing's use of HCL Technologies to provide software engineering services for the 787 program. Since HCL had already been involved with a number of its tier-one suppliers like Rockwell Collins, GE Aviation, and others both for hardware systems and software development, it offered a unique opportunity of synergy across the fragmented supply chain. HCL was designated the preferred software services company for the entire 787 program.

HCL has long had an R&D bent. It developed the first indigenous microcomputer in India in 1978 (the same time as Apple did) and developed the indigenous relational database at the same time as its global IT peers. HCL also says it developed its multiprocessing UNIX-based OS kernel in the late 1980s, a few years ahead of Sun and HP, which of course, commercialized it far more successfully.

Says Sandeep Kishore, Executive Vice President for the Engineering and R&D Services unit at HCL Technologies, "We have done over 750 projects in the aerospace industry covering avionics, engineering design, and extensive testing. We have been involved in all major commercial aerospace programs across the globe in the past decade. Under this umbrella agreement with the Boeing Company, HCL worked as the engineering services partner of choice with multiple sub-tier companies on the 787 program." HCL's services were utilized by these airborne systems suppliers across all the major design elements of the 787 such as common core systems, open systems architecture, and e-enabled architecture.

Some of the key systems that HCL was involved with included electric power generation and distribution systems, remote power distribution systems, air management systems, integrated surveillance systems, display and crew alerting systems, common data networks, and pilot controls.

HCL leveraged its offshore infrastructure and processes to ensure that almost 80 percent of the effort was delivered from its engineering design centers and labs in India.

To help test these systems that were being built all over the world, HCL helped build automated test equipment, load generators, test simulators, and data banks by remotely integrating with the labs of the Tier 1 suppliers during off-peak hours and ensuring optimal utilization of costly test lab infrastructure.

There were several advantages of having HCL across the board working with all its partners for the 787 development for Boeing. First off, it minimized compatibility issues that come from integration of multiple subsystems. It also made sure that HCL cross-leveraged the knowledge and learning on the 787 program across suppliers to speed up the development process.

Says Kishore, "We estimate HCL had some level of involvement and contributed to developing and validating almost 40 percent of airborne software. At peak, over 900 HCL engineers worked with Boeing and 10 of its sub tier partners from the U.S., Europe, and Australia."

Meanwhile, Boeing had a flight test system that needed to be enhanced to test the different versions of the 787. The test system had evolved over several decades and Boeing wanted to migrate the system to a modern architecture that could also potentially be airborne in future versions. HCL assisted in the rehosting of this flight test system.

Another major initiative at Boeing is the Test Operations Center (TOC). The TOC is a 32,000-sq.-ft. facility at the Flight Test Center at Boeing Field. The heart of the TOC is a 2,000-sq.-ft. control room, nicknamed the "bubble," with 6 × 16-ft. screens displaying the status of aircraft in the test fleet. The room brings together engineering, flight operations, and maintenance staff. The TOC is open three shifts a day, with flight tests during daylight hours, ground tests during second shift and maintenance taking place during the third, overnight, shift. In addition to Boeing employees, the TOC is designed to house employees from suppliers and the FAA. HCL played a key role in development of the visibility applications and the database that are part of the Test Operations Center.

In another innovation, HCL also agreed to tie a portion of its compensation for work with the tier-one suppliers to sales of the plane. So, while Boeing has suffered from the delays, HCL is also sharing in some of that pain.

CEO Jim McNerney summarizes: "While we clearly stumbled on the execution, we remain steadfastly confident in the innovative achievements of the airplane and the benefits it will bring to our customers."[26]

As ANA and other charter airlines roll out the 787 on their routes, passengers are finding the cleanest air of any airplane ever built (the air has the same microbial content of outside air with its filters with an efficiency of 99.97 percent) and when they glance out of their 18 5-inch-tall 787 windows, the largest on any commercial plane, and which have electronically adjustable electrochromic dimmers, the production delays are a memory from the past.[27]

In the meantime, Boeing has learned plenty about how to manage a globally distributed technology supply chain.

Chapter 10

Maverick: No Rules. Just Right.

Tom Cruise played United Naval Aviator Lt. Pete Mitchell and flew the F-14A Tomcat with a call sign of Maverick in the 1986 movie *Top Gun*. Unauthorized tower flybys and dialogue like the one below with his instructor made the cocky, macho Maverick an inspiration to many, not just in combat.

Viper	In case some of you are wondering who the best is, they are up here on this plaque. [turns to Maverick]
Viper	Do you think your name will be on that plaque?
Maverick	Yes, sir.
Viper	That's pretty arrogant, considering the company you're in.
Maverick	Yes, sir.
Viper	I like that in a pilot.

The Outback Steakhouse restaurant chain uses a tagline of "No Rules. Just Right." to convey its image of the untamed Australian outback. Fittingly, it is an American invention. The founders had never been to Australia when they started the restaurant, and there certainly is no "bloomin' onion," one of its more popular appetizers, that grows in Australia. One of the founders was quoted as saying, "I might have tried to bring back authentic Australian food, which Americans don't generally like. Our company sells American food and Australian fun." And Australian beer and wine, which does sell to Americans![1]

In technology it certainly helps to break the rules, because "disruption" is a respected term in much of the industry. Breaking rules, however, requires a plan with its own rules and discipline like the kind John Boyd mastered.

"40-Second Boyd"

Peter Fingar, author of many technology books and a frequent speaker at many technology forums, likes to invoke Boyd, a U.S. Air Force fighter pilot trainer and a real-life version of Viper, referenced earlier.

> Did you ever wonder how a smaller, slower, shorter-range, and lower-altitude jet fighter could beat its far more endowed enemy aircraft in a supersonic dogfight? Well, we could turn to the "Father of the F-16," the late John Boyd, who issued a standing challenge to all comers. Starting from a position of disadvantage, he'd have his jet on their tail within 40 seconds, or he'd pay out $40. Stories have it that he never lost. Boyd's maverick lifestyle and unfailing ability to win any dogfight in 40 seconds or less earned him his nickname "40-Second Boyd."
>
> In a dogfight, the F-16 seems to ignore the laws of physical science. It turns energy on and off in a second, and despite its light weight, it can withstand nine times the force of gravity, which enables some serious twisting and rolling. The F-16 allows its pilot to *outmaneuver* the other guy, just as a company would like to be able to outmaneuver its competitors that are bigger, better, stronger, and faster. By being able to sense and respond to the changing competitive environment with great speed, the agile business can confuse its competitors, and by the time they have figured out what happened, the agile business again goes on the offensive in the very environment it had just changed.

Boyd developed a unifying theory of this thing we call *agility*. His key agility concept was that of the *decision cycle* or OODA Loop (Observe, Orient, Decide, Act):

- *Observe:* The lightning-quick collection of *relevant* information about your current environment by means of the *senses* rather than drawn-out data analysis that leads to "analysis paralysis." A huge challenge is picking the right data to focus on in order to avoid information overload.
- *Orient:* The analysis and synthesis of information to form one's current mental perspective. This is the most important step because this is where information is turned into situational awareness, not unlike what's needed in a 3D chess game.
- *Decide:* The determination of a course of action based on one's current mental perspective. Again, avoid analysis paralysis, as a dogfight happens in real time.
- *Act:* The physical carrying out of the decisions. Of course, while your action is taking place, it will change the situation, favorably or unfavorably. Hence the loop repeats, again and again, in real time, as nothing is standing still.

Leading companies are taking this "unsystematic" approach to business innovation and turning it into repeatable, managed business processes. GE calls it CENCOR (calibrate, explore, create, organize, and realize), and it centers on Design for Six Sigma. The Mayo Clinic calls it SPARC (see, plan, act, refine, communicate). Although leading companies and innovation consultants have many innovation process road maps, OODA loops provide the baseline, a unifying theory of agility, for any business innovation process worth its salt.

Maverick CIOs

Chris Murphy, who writes the Global CIO column for *InformationWeek*, gets to observe and write about countless technology executives and says:

"One thing is that 'the rules' are different for every CIO and every company. I've written about a number of what I would consider IT transformations, and what strikes me is the very different approaches they've taken depending on the company. There isn't one playbook, or even one set of rules."

But he does cite three standout examples:

1. *Procter & Gamble's CIO Filippo Passerini.* "The wall in Procter & Gamble's executive conference room, called the Business Sphere, is a twenty-first century monument—built to honor the gods of data. Two white concave screens, about 8 feet tall and 32 feet wide, face each other around an oval conference table, letting as many as 16 executives, led by CEO Bob McDonald, see at once the reality of how P&G is doing around the world."

   But is the data as perfect as its gleaming altar suggests? Is it as real time and comprehensive as this global consumer goods giant would like? No. "We intentionally put the cart before the horse, because it is a way to force change," he says. "For years, companies have tried to gather all the right data, build the high-powered data marts to support it, and only then build decision-support tools to exploit the data. And that approach hasn't worked."

   "Passerini has taken these risks, repeatedly, as he has dramatically changed the role of information technology at P&G, the $79 billion-a-year maker of consumer megabrands from Tide to Pampers to Pringles."[2]
2. *Fedex's CIO Rob Carter.* "More than most CIOs, he insists on pushing business executives to understand technology—as he says, why it's 'too slow and costs too much' to do things like new cross-business integrations and interfaces. Yes, put tech in a business context. But Carter thinks business unit execs must understand even deep concepts such as a services architecture. 'The more we kept this stuff behind the scenes, the less real it was to the business,' he says."
3. *HP's ex-CIO Randy Mott.* "Mott believes in doing fewer projects at a time, finishing each in a shorter time frame, and doing more total projects. He does so by staffing projects up with teams of 10 to 30 people instead of two to 10 people. With that approach, HP cut the typical project time frame from three years down to six months."

Maverick Technology Vendor Executives

"It's the opposite of sticking to your knitting. It's when you shouldn't have stuck to your knitting and you did," says Jeff Bezos, CEO of

Amazon. So he encourages people at Amazon to ask "why not?" when considering whether to launch something new. "It's very fun to have a culture where people are willing to take these leaps. It's the opposite of the 'institutional no.' It's the institutional yes. People at Amazon say, 'We're going to figure out how to do this.' "[3]

Other technology mavericks include Marc Benioff, CEO of Salesforce.com who stands out because he respects leaders from consumer tech even though he sells to enterprises. "We stand on the shoulders of giants like Jerry Yang [founder of Yahoo!], Pierre Omidyar [of eBay], Jeff Bezos [of Amazon], and Biz Stone [of Twitter]."[4] On the other hand, he can be merciless to his fellow enterprise competitors from Microsoft to SAP. Seeing Microsoft reps outside his Dreamforce conference soon after the company had sued his, he told the audience, "I guess you know you've made it in software when Microsoft is protesting your event. Or maybe when they've sued you. This is the greatest thing that's ever happened in my career."

Larry Ellison of Oracle stands out in the industry partly because of his "bad boy" image, but also because of his varied personal interests. His sailing team won the America's Cup. His Katsura House is a 23-acre Japanese-style imperial villa in the Bay area with engineering innovations to make it earthquake-proof in a quake-prone part of the world. He pilots his own planes and has nice taste in cars, like Bentleys. His suits are tailored by Atollini, and his shoes made by the exclusive English store Edward Green. That certainly stands out in the blue jeans and sneaker culture in technology.

Dave Duffield, co-CEO of Workday, is a maverick in tech because he is in many ways the opposite of Ellison—a nice, humble guy. "Dave Duffield has made the big bets on technology, time after time after time. Was first to redesign for RDBMS, at least in my world, at Integral Systems. Bet on Windows GUI and 'client server' at PeopleSoft and literally turned our industry on its head. Bet on models-driven development, objects living in memory to be both the application and the database, and more with Workday [described in Chapter 8]. And he's insisted across the 30-plus years that I've known him on treating his customers and his employees as though they were family. Not a bad track record," says Naomi Bloom, a respected consultant in the human resources world.

Vineet Nayyar, CEO of HCL Technologies (discussed in Chapter 9) has made a name for himself by turning conventional management upside down and adopting what he calls an "Employees First, Customers Second" model. He has also authored a book on that philosophy.[5] Phil Fersht, who analyzes outsourcing firms at Horses for Sources, says Nayyar is "unafraid of not following the crowd, even if his views can be a little 'out there' occasionally." An example of "out there": at a conference in 2010, Nayyar said cloud computing is "bullshit," to the horror of his staff who market HCL's growing commitment to cloud computing. But who said you could carefully script what mavericks say?

William Taylor, co-author of the best-selling book *Mavericks at Work*,[6] lists 50 mavericks on his site.[7] This includes Mark Cuban, who made his riches via various technology media properties and is now owner of an NBA team, appropriately named the Dallas Mavericks. Cuban has been repeatedly fined by the NBA for at least $1,500,000. Taylor also lists Tom Peters, the famous management author, who tells everyone to call on their "saviors-in-waiting," including disgruntled customers, off-the-scope competitors, rogue employees, and fringe suppliers.

Maverick Technology Companies

There are plenty of disruptive vendors across technology categories:

- SaaS vendors like NetSuite and Workday are packaging software, hosting, application management, and upgrades that companies used to buy from multiple vendors.
- Third-party maintenance vendors like Rimini Street are offering cheaper alternatives to maintenance contracts that vendors like SAP and Oracle provide their customers.
- Vendors like Maxroam are offering much more affordable international roaming rates than major carriers like AT&T and T-Mobile.
- Cartridge World offers refilled and remanufactured ink and toner printer cartridges at far lower costs than OEMs like HP and Canon.

- Cloud computing vendors like Amazon provide processing and storage capabilities on a pay-as-you-go basis compared to outsourcers who expect multiyear contracts.

Maverick Technology Product Design

We see it all the time in the auto market—cars so different looking they stand out. As the *Washington Post* wrote about the Scion xD, "Ugly has never been so attractive."[8]

In a crowded technology marketplace, it often helps to have an exotic design or color and stand out from the crowd.

"Cisco's Cius is bulkier than the iPad and has a smaller screen (7 inches wide, compared to the iPad's 9.7). But it packs a number of tricks all of its own, designed to woo business users. The Cius is designed to integrate closely with Cisco's voice and video phone systems, and it can even replace a desktop computer when docked to a new Cisco deskphone, which connects to a monitor, keyboard, and mouse."[9]

In the *Engadget* review of the Motorola Atrix 4G, the reviewer honestly says, "It's hard to fully describe how it works, but it's an unusual combo."[10] It is a smartphone that comes with a dock, which turns it into laptop with a much larger screen and keyboard.

If that is hard to describe, try MusicLites, which is a wireless bulb and speaker in one. It is the collective result of "Osram Sylvania's expertise in LED solutions and Artison's creative dedication to state-of-the-art audio." Oh, and by the way, it also comes with a remote control.

Next we have the Eye-Fi, which brings that combination concept to cameras. It combines SD card and wi-fi so users can upload photos and videos wirelessly from camera to computer without having to physically move the tiny storage card around.

In Chapter 16 we will see the vertical platen of the Lexmark Genesis printer. It towers over most devices on the table or on the floor.

One of the coolest things users saw with the iPad2 was an accessory—its SmartCover, which comes in several different colors. "Built-in magnets draw the Smart Cover to iPad for a perfect fit that not only protects, but also wakes up, stands up, and brightens up your iPad."[11]

Conclusion

In technology, more than any sector, being a maverick is tolerated, even encouraged. Being a maverick does not mean no rules. It means defining those rules to suit your position. Apple has shown time and time again how it breaks other's rules, while defining new ones. In the next section let's look at 10 gutsy moves Apple has made over the past decade.

Case Study: Apple—A Thousand "Nos" and Ten Gutsy "Yeses"

Dan Wood is founder of Karelia Software. His company logo shows two men pumping a handcar—the kind used to maintain rail tracks and resolve other right-of-way issues. The story behind the logo goes back to 2002.

According to *Mac Observer*, "You could see people mouthing the word Watson when Sherlock 3 was demonstrated at MACWORLD New York on July 17. Watson, an Internet services application, was originally released by Karelia Software back in November of 2001 as a complement to Apple's Sherlock search software."[12]

Wood had been called in to Apple prior to MACWORLD and shown Sherlock 3. On his blog he describes his reaction:

> I drove home, gradually realizing what I had just witnessed, and sent off an e-mail to my contact at Apple Developer Relations expressing my unhappy sentiments. An hour later, Steve Jobs called me.
>
> "Here's how I see it," Jobs said—I'm loosely paraphrasing. "You know those handcars, the little machines that people stand on and pump to move along on the train tracks? That's Karelia. Apple is the steam train that owns the tracks." So basically the message was: Get out of the way, kid; this is our market.[13]

Fast-forward to 2011.

Answering a question at the company's shareholder meeting, Apple's leader of iOS development, Scott Forstall, pointed out that Apple has, in the iOS App Store, "created the best economy in software in the history of the planet."[14]

If anything, Apple is now being accused of having allowed too many "handcars on its tracks." Trip Hawkins, once Apple's director of strategy and marketing, left the company in 1982 to found Electronic Arts and other start-ups. He now complains Apple has "overencouraged supply" in its App Store.[15]

The App Store is now considered such a strategic asset that Apple took the unusual step in 2011 of filing "to intervene in the lawsuits filed by patent troll Lodsys against iOS developers. If the court accepts Apple's request, then Lodsys is no longer just challenging a bunch of small developers, but will have to fight Apple's formidable legal team and storehouse of patents."[16]

Time and again Apple has shown willingness and ability to change course. It has mastered the art of being a maverick, challenging rules set by others, and defining its own rules.

Saying No to 1,000 Things

In a 2004 *BusinessWeek* interview, Jobs said, "And it comes from saying no to 1,000 things to make sure we don't get on the wrong track or try to do too much. We're always thinking about new markets we could enter, but it's only by saying no that you can concentrate on the things that are really important."[17]

In his biography of Steve Jobs, Walter Isaacson shows many examples of his amazing ability to cut through the clutter and "say no to 1,000 things." Apple's ability to say yes to plenty of gutsy, even maverick, decisions is what really defines its success.

Let's explore 10 (out of many more) high-risk and high-impact decisions Apple has made.

Popularizing MP3 Singles in Spite of the Music Industry

Think about it—our radio stations have long played popular singles. When we were younger many of us created cassettes of our favorite songs and played them on our Sony Walkmans and in our cars. The music industry preferred to sell us complete albums, then CDs.

While Apple did not invent MP3s, it dragged the music industry with iTunes to make it a big business.

Even today, though, many music executives long for the return of albums. They ignore the "long-tail" phenomenon. Consumers buy many more "one-hit wonders" than they would if they were not unbundled.

Here is an example of such music industry thinking:

> "Historically, the price of an album was five times greater than a single," said [Tom] Silverman [a music executive], who believes setting

the price at a tenth of an album's cost was a mistake and that even $1.29 is too low. "It should've been $1.99, and then we would've seen higher digital album sales because it would've been a bigger discount for buying an album."[18]

Apple persisted. Over 15 billion songs—most of them singles—have been downloaded via iTunes and many more through other channels like Walmart.com.

An Ambitious Phone with Little Previous Phone History

Jobs explained to *Fortune* the rationale behind attempting the iPhone: "We realized that for almost all—maybe all—of future consumer electronics, the primary technology was going to be software. And we were pretty good at software" and "The reason that we were very excited about the phone, beyond that fact that we all hated our phones, was that we didn't see anyone else who could make that kind of contribution. None of the handset manufacturers really are strong in software."[19]

While all that is true, hundreds of other companies could say they are pretty good at software. Was that alone sufficient to make the iPhone so successful? Apple's previous collaboration with Motorola on its ROKR phone just a year prior had not been very successful.

There were many cynics. Verizon President and Chief Operating Officer Dennis Strigl said in 2006, "The iPhone product is something we are happy we aren't the first to market with." Verizon finally started supporting it in 2011.

Prof. Clayton Christensen, who has written treatises on disruptive technologies, was of the view that "Nokia, Samsung, LG, and Sony Ericsson have a lot to lose. You better believe that each of them has a skunk group dedicated to preparing a response to the iPhone. And since the iPhone is a high-end, high-cost product it leaves rivals an opportunity to create a more affordable version that's almost as good."[20]

Turns out the competition was unprepared. "The initial reaction from competitors, or soon-to-be competitors since Apple wasn't really in the game yet, was either shock or laughter. RIM didn't think it was possible to have such a device without it being a power hog. Microsoft's Steve Ballmer laughed at it for not having a physical keyboard."[21]

The iPhone was an audacious move on Apple's part given it was not a telecom insider. The rest, as they say, though, is history.

Signing Exclusive Multi-Year Deals with Carriers

When it introduced the iPhone in 2007, Apple signed exclusive arrangements with Cingular (later rebranded AT&T) for the U.S. market, O2 for the UK market, Orange for France, and other similar deals in other markets. The AT&T arrangement in particular caused Apple significant heartache as that network could not cope with the data demands the iPhone put on a network designed more for voice traffic. There were other customer issues due to outrageous international roaming charges and early termination and upgrade fees.

However, Apple was able to introduce with AT&T innovative features like Visual Voicemail and redefine the customer activation process.

> By allowing users to activate their iPhone from the comfort of their home or office, Apple is redefining the phone-buying experience.
>
> No more waiting at the store while a bumbling employee tries to get your phone registered with the carrier. No more hassling with said bumbling employee to transfer your old phone numbers to your new phone. Apple iTunes will walk you through a step-by-step process to do it all yourself![22]

The AT&T relationship came at a significant price. TechCrunch summarized: "I understand why Apple went exclusively with AT&T at first (though it had first offered the device to Verizon, which turned it down)—it got a pretty sweet deal, and was able to use it to put it in a position of power over the entire industry. And I even understand why they re-upped the first time—to get an even sweeter deal (the subsidy from AT&T for each phone sold). But now AT&T is a liability for Apple that will inhibit its huge potential for growth in the U.S."[23]

To its credit, Apple stuck with AT&T and continued even further, even as it brought Verizon in as an option.

Contract Manufacturing Arrangement with Foxconn

In the 1990s, Apple made a strategic decision to deemphasize its own manufacturing. Metrics such as inventory turns did not look impressive especially when compared to Dell, which was then a paragon of efficiency. Plants in Ireland, Singapore, and the United States were outsourced. Tim Cook, now Apple's Chief Executive Officer, gets credit for that supply chain transition. Foxconn, the Taiwanese contract manufacturer, benefitted in particular from the outsourcing.

Apple has stayed loyal to Foxconn through multiple releases of iPods, iPhones, and iPads in spite of sweatshop accusations, employee suicides, and explosions in its Chinese facilities. Most companies like to diversify between company-owned and outsourced plants, and certainly across geographies. Apple has been willing to take the risk of putting its eggs in the Foxconn basket and taking plenty of jabs from China bashers (although Foxconn has recently been looking at a significant diversification to Brazil).

Over 200 million iOS devices sold to date, most assembled by Foxconn, justify the loyalty. Its regular audits of the supplier appear to confirm what Joel Johnson wrote in *Wired* magazine. This was after a visit to Foxconn facilities after the spate of employee suicides. "But the work itself isn't inhumane—unless you consider a repetitive, exhausting, and alienating workplace over which you have no influence or authority to be inhumane. And that would pretty much describe every single manufacturing or burger-flipping job ever."[24]

Investing in a Large Retail Operation

When Apple opened its first retail store in 2001, an observer commented, "I give them two years before they're turning out the lights on a very painful and expensive mistake."[25] Apple dealers were not happy. One of them said, "Apple might do just well enough to really hurt our business. They've also done a poor job telling us how sales and service will work when an Apple Store opens nearby."[26] There were others who questioned the timing in 2001 with the tech industry in a severe downturn. They drew parallels to PC maker Gateway's struggling retail effort.

Apple, of course, has gone from strength to strength with its retail operations, as we describe further in Chapter 12. Over 230 million visitors entered one of Apple's over 300 stores around the world in 2010. In the ultimate compliment, Apple's SVP of Retail, Ron Johnson, was recruited in the summer of 2011 to become CEO of J.C. Penney. A decade after the first store was opened in Virginia, the Genius Bar at the stores really looks like a Genius move.

The iPad Decision after Many Industry Tablet Failures

When Apple introduced the iPad in 2010, it tried to reverse "more than two decades of (industry) attempts to get tablets right—none of which really succeeded, and some of which failed on a monumental scale."[27]

John Dean, former CIO of Steelcase and a big user of tablets himself for almost a decade, explained the challenge: "If I could get 15 minutes to spend with anyone, I could convert them to using a tablet. But I would only target those that were doing things where a tablet offered a better way." "When you mess with the physical interaction aspect of a technology, the change can be traumatic. We are stuck in a keyboard/text/formal content world with fingers (or thumbs) as the means of interaction."[28]

Others were harsher. A Dell executive said, "Apple (iPad) is great if you've got a lot of money and live on an island. It's not so great if you have to exist in a diverse, open, connected enterprise; simple things become quite complex."[29]

By June 2011, just 14 months after the iPad was introduced, Apple announced it had sold 25 million units and it had pretty much locked out competition from over 100 tablets that were launched in the wake of the iPad success.

Challenging Amazon in eBooks

Amazon has over the past decade become a dominant player in the book market. As eBooks started to grow in popularity over several years, Amazon had driven the market toward a $9.99 price point. At the iPad launch, Walt Mossberg of the *Wall Street Journal* asked Steve Jobs why people would pay more for the same book on an iPad. Jobs' response was "The prices will be the same."[30]

Apple had no intention of keeping prices that low, even though it was a trivial player in the book market compared to Amazon. Instead, it signaled to publishers it would accept an agency model that would give them more control over pricing. That precipitated a showdown between publishers and Amazon. Amazon ended up ceding price control to the publishers.

Venturebeat wrote, "The ten-dollar e-book may soon be gone, replaced by the fifteen-dollar eBook."

The Mind Games with Microsoft and Adobe

Starting in 2006, Apple ran the Mac versus PC commercial series.[31] It was a series that humiliated Microsoft across 66 commercials. Given the earlier statement that Apple is good at software, it could have tried to develop its own version of Microsoft Office. Its repeated badgering in that series could have annoyed Microsoft to give up on the Apple market.

The reality is Microsoft's continued development of Office products for the Mac has actually made transitions away from the PC easier for many. Phil Fersht, founder of the analyst firm Horses for Sources, says, "We have our whole firm on Macs now—we would likely have never considered that scenario if the Office suite wasn't available."

The attack on Adobe was even less subtle. Wrote Jobs: "Flash was created during the PC era—for PCs and mice. Flash is a successful business for Adobe, and we can understand why they want to push it beyond PCs. But the mobile era is about low power devices, touch interfaces, and open web standards—all areas where Flash falls short." And "New open standards created in the mobile era, such as HTML5, will win on mobile devices (and PCs too). Perhaps Adobe should focus more on creating great HTML5 tools for the future, and less on criticizing Apple for leaving the past behind."[32]

In November 2011, Adobe took his advice to heart and announced they would deemphasize Flash for mobile platforms. Apple, long called a "closed vendor," had actually driven the market toward HTML5, an open standard. How maverick is that?

Large-Scale Sourcing of Strategic Components

While many technology companies use cash reserves to make acquisitions, Apple has on many occasions used large sums of money to secure strategic components. In 2005, "Apple reached long-term supply agreements with a number of memory supply companies including Hynix, Intel, Micro, Samsung Electronics, and Toshiba. Apple will prepay up to $1.25 billion for flash memory over the next three months. The agreements secure a supply of NAND flash memory through 2010."[33]

In 2011 Apple announced, "During the September and December quarters, we executed long-term supply agreements with three vendors through which we expect to spend a total of approximately $3.9 billion in inventory component prepayment and capital expenditures over a two-year period."

Apple would not reveal the components or the suppliers, but such large commitments not only get them priority in tight markets but also freeze out competitors.

Doubling Down on the MobileMe Failure

At Apple's developer conference in June 2011, Jobs showed his sense of humor as he announced the iCloud service with this quip, "You might ask, why should I believe them? They are the ones who brought me MobileMe."

MobileMe was not one of Apple's best products. *Wired* magazine wrote: "Tech observers agree that MobileMe has been one of Apple's most embarrassingly flawed products, thanks to its extremely buggy launch and limited functionality. MobileMe was itself a 2008 rebranding of .Mac, which began its life in 2000 as iDisk."[34]

In fact, the industry view was that if Apple had a chink in its armor it was in web services—Amazon, Google, Facebook, and several others were far superior. In the months prior to the iCloud announcement,

Amazon and Google beat Apple to the punch with their own versions of streaming music offerings.

So what does Apple do? Invests in a massive 500,000-square-foot data center and offers a Match feature with iCloud. "If you have to upload your whole library (as you would with the Amazon and Google offerings), that could take weeks. If we're scanning and matching, we don't have to upload them. They're in the cloud. It takes just minutes. Not weeks."[35]

The Match feature came at $25 a year—a pittance for those with large music libraries. The time saving was the killer feature. Not to mention, the quality of the music is much better. It is at 256-Kbps iTunes Plus quality, even if your original copy was of lower quality.[36]

It could be saying no to 1,000 things or it could be saying yes to 10 gutsy decisions, or it could be what Marc Benioff, who did a stint at Apple before founding his own very successful Salesforce.com says, "Steve Jobs is the best technology leader the world has ever seen."

Certainly, most would agree Jobs was a maverick. And the "Top Gun" of the industry!

Chapter 11

Malleable: Business Model Innovation

"DOA (Dead on Arrival)!" exclaimed Jonny Evans in *Computerworld* about tablet competition as Apple announced version two of its iPad.[1] Jason Perlow ran a blog on ZDNet a few weeks later with a picture of the HP TouchPad with the same acronym "DOA" in bold and red.

Perlow said, "The only way the iOS tablet ecosystem can be disrupted by a competitor is to come in substantially cheaper. I've said this as well about Android tablets—without a comparable ecosystem, you have to come in as more value-priced."[2]

He was prophetic. Just a few weeks later HP announced it was killing the product and announced a fire sale with some versions going at $99. Later in 2011, Amazon announced its tablet, Fire, at $199. Definitely value-priced!

Samsung and other tablet vendors may disagree—indeed, they should be offended with the DOA designation, but as we have described in other chapters, Apple's efficiency in manufacturing and logistics allows it to compete ruthlessly on price while continuing to be extremely profitable.

It was fascinating to watch, in contrast, when the iPad was introduced, Apple actually managed to convince book publishers to *increase* e-book pricing beyond the market standard being set by Amazon.

In technology circles, we look for competitive advantage via form/factor and feature/function. Mark Johnson of the consulting firm Innosight (which he co-founded with Clayton Christensen, well known for his work on disruptive technologies) says, "It's far harder for an incumbent to fight back against a business model innovation than it is for them to match and raise the stakes on a technology innovation."[3]

Repeatedly, Apple has shown a willingness and ability to compete on "business model innovation." Amazon does the same—it lost money on many e-books at $9.99 as it tried to build customer loyalty around a new way of reading books on Kindles. It has long subsidized shipping of physical books and other items and it offers customers Prime membership, which for $79 a year (in the United States; prices vary in other countries) includes two-day shipping. This is a great deal for frequent customers.

Competing on Business Models

The strategy consultants at Boston Consulting Group define business model as having two essential elements, the value proposition and the operating model.[4]

The value proposition "reflects explicit choices along the following three dimensions:

- Target Segments. Which customers do we choose to serve? Which of their needs do we seek to address?
- Product or Service Offering. What are we offering the customers to satisfy their needs?
- Revenue Model. How are we compensated for our offering?"

The operating model "captures the business's choices in the following three critical areas:

- Value Chain. How are we configured to deliver on customer demand? What do we do in-house? What do we outsource?
- Cost Model. How do we configure our assets and costs to deliver on our value proposition profitably?
- Organization. How do we deploy and develop our people to sustain and enhance our competitive advantage?"

Technology is allowing industry after industry to look at new business models. GM used its OnStar navigation and emergency services as a car selling feature. As we described in Chapter 1, it is now unbundling the functionality and packaging it into a rearview mirror that allows anyone—even with a Ford or a Toyota—to sign up at $18.95 per month. We mentioned in Chapter 6 how Virgin America allows you to upgrade for a fee while you are on the plane. Try doing that on other airlines, with their policies that cut off upgrades hours or days prior to a flight.

In consumer markets, mobile payments are increasingly allowing for micro-payments and attracting new consumers "at the bottom of the pyramid." Of course, they also allow targeting more affluent consumers. Soon you should be able to activate ski insurance for just the day you ski, via your cell phone. Evernote, with apps that help users document and remember trivial and important things, epitomizes the "freemium" business model. Many of its 40,000 new users a day happily upgrade to its paid service after trying out its free products. "Virtual Currency" is growing in popularity, as Zynga shows with its ZCoins and Facebook with its credits.

Don Tapscott, the leading management thinker, puts it succinctly: "There are two kinds of business models: those that have been disrupted by technology, and those that have yet to be."[5]

Let's look next at some markets that have seen seismic, technology-driven shifts in business models, and others poised to be similarly affected.

Book Publishing

Amazon accounted for an estimated 80 percent of all electronic-book sales, and $9.99 seemed to be established as the price of an e-book. Publishers were panicked. David Young, the chairman and CEO of

Hachette Book Group USA, said, "The big concern—and it's a massive concern—is the $9.99 pricing point. If it's allowed to take hold in the consumer's mind that a book is worth 10 bucks, to my mind it's game over for this business."[6]

Apple agreed to an agency business model as it launched its iPad in early 2010.

"Under those agreements, the publishers will set consumer prices for each book, and Apple will serve as an agent and take a 30 percent commission. E-book editions of most newly released adult general fiction and nonfiction will cost $12.99 to $14.99."[7]

A few days after the iPad was introduced, a public spat ensued between Amazon and several of the major publishers like Macmillan. John Sargent, CEO of Macmillan, described the disagreement in an open letter:

> This past Thursday I met with Amazon in Seattle. I gave them our proposal for new terms of sale for e books under the agency model, which will become effective in early March. In addition, I told them they could stay with their old terms of sale, but that this would involve extensive and deep windowing (delaying of ebook versions compared to hardback versions) of titles. By the time I arrived back in New York late yesterday afternoon they informed me that they were taking all our books off the Kindle site, and off Amazon. The books will continue to be available on Amazon.com through third parties.[8]

By Sunday, Amazon announced on its website: "We will have to capitulate and accept Macmillan's terms because Macmillan has a monopoly over their own titles, and we will want to offer them to you even at prices we believe are needlessly high for e-books."

Needlessly high? Apple can be credited—or blamed—for changing the business model.

The Mobile Phone Industry

"The more things change, the more they remain the same." That expression could certainly apply to the mobile services marketplace. Prior to Apple introducing the iPhone, carriers paid device manufacturers just enough for "planned obsolescence," and subsidized those devices to consumers. This allowed them to lock customers into another long-term contract with the new phone. There was little device loyalty. In fact, carriers like Verizon had the marketing slogan "new phone every two years."

The iPhone is a platform with frequent software updates and an application ecosystem. So it has a much longer brand commitment. The device used to be the commodity in the equation. Now as Apple expands its carrier choices—AT&T, Verizon, and Sprint in the United States—the mobile service becomes the commodity in this changing business model.

"According to Paul Roth, AT&T's president of marketing, the carrier is exploring new products and services—like mobile banking—that take advantage of the iPhone's capabilities. "We're thinking about the market differently," Roth says. In other words, the very development that wireless carriers feared for so long may prove to be exactly what they need. It took Steve Jobs to show them that."[9]

The telecom industry has been a fiendishly complex one for Apple to navigate. Working with 200 carriers around the world, each of which is quasi-regulated, is a considerable logistical challenge. Many, like AT&T in the United States, were just not prepared for the heavy data demands the iPhone put on their network, which previously had mostly been geared for voice traffic.

The bigger challenge was that most of the telecoms wanted to continue with older business models, even around the innovative new phone. Two problematic areas, in particular, were long-term contracts (and related early termination fees) and hefty international roaming charges.

> **Exhibit A:** "Ross had no AT&T service after the nearby cell-phone tower went down and the other towers weren't working. Even still, AT&T wouldn't let him leave service without paying an early termination fee, despite the fact that it could be four months before the towers were repaired. That means four months more without service while still getting a monthly bill."[10]

> **Exhibit B: Comment from a customer named Mike:** "You can imagine, I was totally floored. I had traveled all over India and Europe for weeks at a time and only incurred a few hundred dollars. Here I am in Canada with an $800 data bill. Pretty shocking.

> Here is something to keep in mind when you are traveling to Canada planning to use your iPhone: International voice roaming in Canada is $0.59/minute. High, but pretty much to be expected. International data roaming is $.0195/KB. That is about $20/MB. 1 MB is about the size of a small digital photo downloaded in an email attachment. No wonder my data bill was that high."[11]

While Ross and Mike may seem like extreme examples, they are actually pretty mainstream. *Consumer Reports* called AT&T its "lowest-scoring cell-phone carrier in the U.S.," according to a satisfaction survey of 58,000 and it "found that iPhone owners were much less satisfied with their carrier and rated data service (web and e-mail) lower than owners of smart phones on other carriers that, like the iPhone, have a host of apps to encourage heavy data use."[12]

Apple has moved away from its exclusivity with AT&T in the U.S. market, but Verizon's service was announced by *Wired* magazine as "Any smartphone customer who uses an 'extraordinary amount of data' will see a slowdown in their data-transfer speeds for the remainder of the month and the next billing cycle. It's a bit of a bait-and-switch. One of Verizon's selling points for its version of the iPhone is that it would come with an unlimited data plan—a marked contrast to AT&T, which eliminated its unlimited data plans last year."[13]

Sprint, introduced as a new iPhone carrier in late 2011, has been running commercials targeting the data limitations imposed by its competitors.

Apple has smartly been beefing up its iPod to give customers most of the iPhone functionality without a carrier contract. Karin Morton, a consultant from Brazil who travels extensively, says, "I make calls using an Apple iPod Touch with the iPhone Stereo Headset, which has a microphone in its cord and costs about $30. Most people do not realize the iTouch does most things the iPhone does and you do not need to sign up for mobile voice and data plans. If you can get on wi-fi, the free Skype application allows you to call and SMS anywhere."

It is a miracle Apple has succeeded with its iPhone saddled with its carriers and their business models. It has, however, opened the door for Google in particular to allow other device manufacturers and carriers to rally around its Android platform. According to Gartner, Android's share of the smartphone market has almost quadrupled over the last year, and now stands at more than double that of Apple's iOS.[14]

Many of those Android-focused carriers are competing with different business models. MetroPCS, which touts its no-contract policies, announced in late 2010, "This holiday season, we're giving customers what they've wished for—an affordable Android smartphone with a full spectrum of applications and unlimited services to do more like never before." Virgin Mobile introduced a similar offering: "With a $149 initial investment and then an ongoing cost of less than $1 per day, someone can have a basic, but useful, smartphone in the U.S., with the flexibility of upgrading to a better phone or different carrier at any point in time."[15]

Dennis Howlett, a ZDNet blogger based in Spain notes, "As a regular traveler between Europe and the United States, telco roaming charges and particularly the egregious data roaming costs charged by the mainstream telcos represent a significant cost and cause of deep resentment. Or at least they did up until my last trip. Long-time colleague Zoli Erdos told me about Virgin Mobile USA, which offers a 'no-contract' deal that includes unlimited data, wi-fi connectivity, plus 300 minutes of talk time for $25 per month. No need for a U.S. address or credit card—just pay as you go—and top up each month or as travel schedules dictate. That's a bargain in anyone's language, which I grabbed with both hands, courtesy of Best Buy. I also took the opportunity to try an Android-powered LG Optimus device that is updated over the wire rather than relying on a SIM chip. Five minutes spent verifying that Virgin does indeed deliver what it says on the tin and another five minutes for the Best Buy representative to activate the service and I was good to go. In an age of instant gratification, that scores highly as the kind of service and provisioning I want. I can top up directly from the Virgin Mobile website. No need to find top up cards or call in. How good is that?

"The Virgin service runs off the Sprint network backbone. While some say Sprint's coverage is spotty, I had no such problems in either the Bay Area or Orlando. Looking at the Sprint coverage map, I should be fine for all the major cities I visit. The Optimus gives me a full day of juice, which I recharge via my laptop's USB connection. It isn't the most advanced smartphone in the world but is more than adequate for my needs. Heck—it even encouraged me to make a few calls, something

I am normally fearful of doing when traveling. At $200 plus tax the Optimus was a luxury treat that will repay itself in weeks rather than years. This is how all telcos should operate: fast, simple, and an easy service to both understand and consume."

Microsoft, which has acquired Skype, and is working with Nokia, has a huge opportunity to target the Mortons and Howletts of the world looking for reasonably priced roaming functionality.

Business Services

Sun Print Management based in Holiday, Florida, is one of a growing number of managed services providers. It provides a service that manages all the onsite maintenance and help-desk support for your laser printer fleet. Customers are charged not for the individual products that they order, but for the number of pages they print. Each printer keeps a running page count. Page counts are collected electronically and transmitted to Sun Print. Depending on volume, some customers pay as little as a penny a page. Again, it's the model—paying for outcomes and at small units of consumption—that is innovative.

Cognizant, the outsourcing firm, has been offering what it calls BPaaS (business processes as a service) for more complex areas than print services, but using a business model that's as easy to understand and consume.

Francisco D'Souza, CEO of Cognizant, explains, "The global services industry is undergoing deep structural changes that are reshaping both the demand and supply sides of the equation. Through wage arbitrage, economies of scale, and productivity improvements, global services providers have, over the last two decades, delivered significant savings to companies across nearly every industry by reducing the unit cost of many operational business and technology processes."

He continues,

> Many companies that have experienced these benefits are now asking: What's next? Our belief is that we are at the very beginning of a profound shift in business technology, one that will significantly reinvent how work is performed within the enterprise. Talent shortages

abound across industries. For example, life sciences companies and healthcare providers are continuously looking for pharmacists and R&D clinicians, as well as doctors. Insurers need more actuaries—it's a dying art form in the U.S. Manufacturers require supply chain experts with a keen understanding of ever-changing energy costs, regional regulations, security challenges, etc. Once obtained and trained, these knowledge workers perform complex activities that are core to the company brand and mission. However, they can reside anywhere, inside or outside the corporate firewall, and need to work collaboratively and have access to insights in real time. What's needed is a new class of service to automate and elevate the quality of work performed virtually across the extended enterprise. Cognizant and other service providers are integrating software platforms, business process work, consulting, and infrastructure to provide end-to-end business solutions that power more dynamic knowledge work in ways that significantly impact the core business of companies across industries.

We are implementing industry-specific and horizontal platforms in areas such as sales and marketing, order management, business intelligence, and clinical data management. In addition to new delivery models, this opens up innovative new commercial models. Rather than pricing services in traditional FTE (full-time equivalent) form, we are building commercial models aligned with business process outputs. We are also increasingly sharing risk and reward by collaborating with client business decision makers to meet or exceed pre-set business performance targets.

These new process-centric solutions deliver cost savings, help clients focus on what is truly core to the business, and leverage new technologies to help our clients work more productively. This is already unlocking new levels of value, and we see an exciting future as companies increasingly embrace these new service models.

Analytics-Driven Business Models

Later we will see in the Valence Health case study how new businesses are mining the mountains of data increasingly available. We already live in a world of "big data," which will soon look tiny as billions of smart devices collect our location, health, energy usage, cooking, and laundry pattern data and with sensors in the field collecting additional commercial data.

These monstrous volumes of data are calling for a new generation of analytical tools. One of those tools is the open-source tool, Hadoop. Per its website:

> The Apache Hadoop software library is a framework that allows for the distributed processing of large data sets across clusters of computers using a simple programming model. It is designed to scale up from single servers to thousands of machines, each offering local computation and storage. Rather than rely on hardware to deliver high availability, the library itself is designed to detect and handle failures at the application layer, so delivering a highly available service on top of a cluster of computers, each of which may be prone to failures.[16]

Hadoop is based on concepts developed at Google and made increasingly robust by other web companies like Yahoo! There are several versions of Hadoop, but a start-up called Cloudera says it has the best Hadoop distribution and it makes money by providing customers services and support around it.

Charles Zedlewski, Cloudera's VP of Products, explains some of the use cases of Hadoop and other "big data" tools:

> The novel capabilities of systems like these have enabled all kinds of analytic scenarios that weren't previously possible.
>
> - Electronics manufacturers can store and analyze usage data from the devices they ship so as to understand product issues in the field.
> - Financial institutions are constructing more elaborate fraud prevention models.
> - Pharmaceutical companies are cleansing and processing vast amounts of genomics data.
> - Retailers are building machine learning models that predict individual customers' buying preferences and make recommendations.
> - Media companies are analyzing viewer data to optimize the content they present.
>
> The consistent theme with these stories has been companies analyzing or processing an order of magnitude more data than they have in the past in order to derive some insight that previously was not attainable. Very often these insights are leading not just to cost savings but new revenues.

The Music Industry

It's tough enough to develop a dominant business model in your own industry as Apple has with its iPads. Try bringing along the music industry with a single song model as iTunes did starting in 2003.

The music industry had little incentive to change: "Between the demise of the 45-single and the rise of CDs, no format existed to sell single songs. That was the honeymoon era for the recording industry, which reveled in selling complete albums to consumers at a handsome price."[17]

Actually, with peer-to-peer (P2P) networks such as Kazaa and Napster, you had been able to download music files for years prior to iTunes. Not that there was a business model in that for the music industry. Napster was sued and shut down and thousands of users were threatened with legal action ruining the brand of the music industry. Apple's big innovation was to have consumers pay for the music. With over 15 billion iTunes song downloads later, many in the music industry are now resentful at how powerful Apple has become.

Apple has continued to innovate both the technology and business models behind its iPods and iTunes. iPods have evolved to a point where they are almost like iPhones—without the hefty monthly payment to a telephone company. In the process they destroyed Microsoft Zune and many a competing MP3 player.

Apple, however, found audiophiles still preferred buying CDs and ripping music from them to downloading from the iTunes store. So, in 2007 it announced with EMI Music a 256 kbps AAC codec "resulting in audio quality indistinguishable from the original recording."[18] That came at a higher price—$1.29 per single compared to $.99, but in the process Apple convinced the studio to make it DRM-free—free of the restrictions music companies levied like a limit of number of devices you could play the music on. Over the next few years, Apple convinced other labels to offer similar higher-quality, DRM-free, iTunes Plus tunes.

In 2011, as Amazon and Google beat it to the punch in offering "music in the cloud," Apple was able to pull its usual "one more thing" as it announced its iCloud. It unveiled the iTunes Match feature. "We have 18 million songs in the music store. Our software will scan what you have, the stuff you've ripped, and figure out if there's a match," Jobs said. [19]

The Apple website added, "And all the music iTunes matches plays back at 256-Kbps iTunes Plus quality—even if your original copy was of lower quality."[20]

While some said the feature was legalizing pirated tracks, in many ways it was a bonus for the music industry. "Apple will split the $25 service fee with record companies and music publishers, giving 70 percent to the music industry and keeping 30 percent for itself. If the service becomes popular (which it very well may, considering the convenience with which it allows music to be shared among Apple's many popular devices), it has the potential to significantly help a music industry that has been under serious financial hardship since music became easily available digitally. Jeff Price, chief executive of the music distributor TuneCore Inc., said that iTunes Match has the potential to 'reset the odometer' for the music industry."[21]

Many Technology Vendors as Laggards

Even as technology companies like Apple and Cognizant, shown above, have been open-minded about business models, way too many technology vendors cling to their older models. Indeed, a UK government agency report blares as its title "Government and IT: A Recipe for Rip-Offs."[22] While you can argue about efficiency in the public sector, in general, larger technology vendors have been loath to change their business models. Here are a few examples:

- IBM announced "On-Demand" services in 2001 in response to growing customer clamor that IT had become too fixed an expense. It isn't IBM, however, but cloud-computing vendors like Amazon and Salesforce.com that are delivering small, bite-sized units of technology procurement and provisioning to make IT spend more variable.
- Major telcos such as Verizon, AT&T, and Sprint sued Vonage, claiming prior art involving VoIP dating back to the 1990s. In spite of claiming prior art, they were not very interested in, and then not

very successful at, offering VoIP to their customers. In fact, many continue to fight providers like Skype from offering VoIP on their mobile networks.

- Accenture (in a predecessor entity, Arthur Andersen) pioneered a software development center in the Philippines in the mid-1980s. It did not scale that low-cost pool for years and did so years later only in response to the global delivery model that firms from India and younger U.S. companies, such as Cognizant, showcased to Western customers.
- Larger software vendors like SAP and Oracle are years late in offering their own version of Software-as-a-Service (SaaS). In fact, it could be said that Larry Ellison, CEO of Oracle, smartly invested in two SaaS start-ups—Salesforce.com and NetSuite—because he realized starting them as projects within Oracle would not generate much success.

Conclusion

Technology-elite companies like Apple and Amazon have shown that creativity in pricing and efficiency in costing are as important as good product design and logistics. Technology is allowing for new models of "freemium," "from the bottom of the pyramid," and other options. In the next section, we see how a young company, Valence Health, using sophisticated analytical technology and a new business model, is building a viable venture.

Case Study: Valence Health

The average of healthcare cost in the U.S. was $7,290 a person in 2007. This was nearly two and a half times the Organisation for Economic Co-operation and Development average of $2,984.[23] Even with that much spending, the U.S. life expectancy, child mortality, and other health metrics are significantly worse than those of other developed countries. And, to accentuate the problem, the number of medically uninsured in the United States grew from 39.8 million in 2001 to 46.3 million in 2008.[24] That's almost 15 percent of the country uncovered in spite of so much spending.

Lost in the politics and rhetoric about the healthcare reform plan the current administration managed to pass and has been fighting to preserve (and which critics deride as "Obamacare") is a stark reality. The U.S. healthcare system is in need of some fresh thinking.

Valence Health is bringing in some of that thinking. It was founded in 1996 by two entrepreneurs, Philip Kamp and Todd Stockard, who both worked in the managed healthcare consulting practice of the accounting firm PriceWaterhouseCoopers.

Stockard points to three flaws in the current healthcare system:

1. Fee-for-service reimbursement—For the vast majority of care provided today, a fee is paid for a service. So, patients largely do not care how many services they require and the costs to provide such care. As a result, there are no incentives in place for healthcare providers to focus on improving the quality of care in order to bring costs down. Providers also have the incentive to do more services because of the fee-for-service compensation model.
2. Lack of population management—Historically, hospitals and physicians have provided care for the patients that present themselves with symptoms and/or an existing medical problem. No one party is responsible for the management of the health of the population with a focus on keeping people healthy.
3. Inability to measure quality of care—Lack of comprehensive data across healthcare providers and disagreement as to how to measure

> quality has created a void in measuring the quality of care and outcomes by individual providers. In many cases, there is lack of evidence as to the impact of different treatment patterns.

Started as a services firm, Valence has broadened its product portfolio, leveraging increasingly powerful analytical tools and opportunities opened up by the reform legislation. Stockard, who graduated from Princeton with a degree in economics, jokingly says he missed his calling. As a young man, he was deep into baseball statistics, and he has carried that analytical bent into this venture.

Nearly 15 years later, with a staff of 120, Valence Health is a turnkey HMO administering the financial, actuarial, data analysis, claims payments, customer service, and medical management of many provider-sponsored health plans across the United States.

Valence also has a clinical integration practice that works with non-risk-assuming groups of doctors and hospitals, giving them the tools to become an integrated system and allowing them to collectively negotiate enhanced reimbursements from healthcare plans.

Stockard elaborates:

> A few years ago, the Federal Trade Commission (FTC) said to doctors and hospitals, "You can't collectively negotiate with health plans unless you're either assuming financial risk or you're clinically integrated." They defined that as creating care guidelines, collecting data and measuring performance against those guidelines. While the evidence-based guidelines are out there, the challenge for these doctors and hospitals is how to collect data from disparate data sources to measure compliance against those guidelines. The health plans won't provide it because it would be used to negotiate against them.
>
> Insurance companies have always been in control of the provider community. They have had the data. Those with the information always win. Historically, independent physicians and hospitals who have entered into risk/insurance relationships with payors have not fared well. The financial incentive for individual doctors continued to be focused on providing more services. Our business is focused on providing information to physicians and hospitals in order to level the playing field.

One of the new concepts introduced by the reform legislation is an Accountable Care Organization (ACO). Explains a report from the National Institute for Health Care Reform:

> To slow medical care cost growth and maintain or improve health care quality, policy makers, researchers, and practitioners have proposed the development of accountable care organizations (ACOs). While there is no single, well-accepted definition of ACOs, there is general agreement that ACOs will constitute groups of providers—physicians, other clinicians, hospitals, or other providers—that together provide care and share accountability for the cost and quality of care for a population of patients. Payers would contract with ACOs to care for a defined group of patients, using financial incentives to encourage ACOs to meet cost and quality goals. Ultimately, policy makers hope that ACOs will improve outcomes and reduce overuse of medical care.[25]

A key concept behind ACOs is clinical integration (CI). Since buyers of healthcare are willing to pay for better quality, provider networks that can define, measure, and demonstrate improved outcomes and add value should be rewarded. A vehicle to achieve these goals for a given group of independent physicians and a hospital system serving a specific population is a clinically integrated network.

Says Stockard, "CI is the process by which hospitals and physicians are able to develop, analyze, and manage patient-centric care, in an effort to produce improved quality, achieve better outcomes, and earn increased reimbursement from payors in exchange for lowering the overall cost of treatment. Valence helps over 10,000 independent physicians use economies of scale that come with clinical integration—even when they lack a common healthcare information platform."

Stockard continues, "We can provide software technology that can pull meaningful data from the disparate information systems of various healthcare providers, and can do so without the personal involvement of physicians or their staff, or the need to source any such data from payors."

Valence created a proprietary tool that accesses a medical community's billing system and uses tools from the analytical vendor, SAS, to scrub the data, apply appropriate clinical care guidelines, and then pass information back to physicians.

Essentially, Valence is providing the benefits of an electronic medical record (EMR), allowing independent practices to see what is happening with a patient across providers. Using a chronic sinusitis guideline as an example, Stockard says, "if somebody shows up in a primary care office three times with that diagnosis, and the guideline says they need an ear, nose, and throat (ENT) referral and a CT scan, we now have the data from everyone in town and can look for that patient at the ENT encounter or look at the radiologist data to see if the CT scan happened."

Additionally, Stockard points out that Valence can now provide alerts about patients before they visit a practice, so doctors have the information they need to ensure compliance with care guidelines. For one client, Valence used SAS Analytics to mine patient data to let doctors know which children needed certain immunizations. They provided doctors with a registry that was integrated with an interactive voice response (IVR) system to make outbound calls to such patients. Stockard continues,

> We've turned our service from a retrospective view of where mistakes might have occurred to a proactive alert system that helps keep the population healthy, When we started, our vision was that health-care providers needed to take control of their destinies. U.S. health-care reform is saying the same thing 15 years later. The concept of accountable care organizations (ACO) and pushing more accountability back to the providers is what our business model is all about.

Valence's advanced patient management tools provide clinical personnel the necessary information to manage the patient through the care continuum.

Seven disease-state modules are utilized:

1. Congestive heart failure (CHF)
2. Chronic obstructive pulmonary disease (COPD)
3. Type 2 diabetes (DM-2)
4. Asthma (including a specific pediatric asthma program)
5. End-stage renal disease (ERSD)

6. Anti-coagulation therapy
7. Coronary artery disease (CAD) Secondary Prevention

In addition to the disease-state modules above, Valence also has effective programs for high-risk pregnancy and catastrophic case management.

The company's consultants utilize its Disease Manager Tool to track the continuum of patient care and assist clients in actively managing high-risk populations. Certified case managers and disease managers interact with patients to foster acceptance of and compliance with recommended treatment and facilitate best outcomes.

The approach is working and Stockard notes, "The ROI for doctors is enormous. As a result of being clinically integrated through our process, physicians have been able to negotiate rate increases of between 15 and 20 percent with health plans. Before, individual doctors had no leverage in negotiations. In one region, doctors were getting 110 percent of Medicare amounts prior to clinical integration. Once they cleared the FTC hurdle and negotiated together, they got 130 percent of Medicare. Clinical integration facilitates the assumption of financial risk, allows doctors to compete for more market share, and also provides patients with better access to more informed care."

The Twists and Turns

Kaiser Health News cautions that "creating ACOs requires hospitals and doctors to work closely together and to share financial risk as well as potential profits."

"There are political and cultural challenges," said Elliot Fisher, the Dartmouth health expert who helped devise the concept. "Many doctors still hold autonomy as a high attribute and many don't want to be bossed around or be employees."[26]

Stockard says Valence actually provides an alternative.

> Hospitals and health systems are in the business of filling beds. Physicians fill beds and in order to continue to fill beds, hospitals have been looking at a number of strategies to ensure physician "loyalty." One option for hospitals has been to buy physician practices. If typical primary

> care practices are valued at two to four times income, a three-physician practice could cost anywhere from $1.2 to $2.5 million. They are also now in the business of physician practice management, an area where many of them have zero experience.
>
> We have presented a second, more cost-effective physician alignment option to our hospital clients. By providing technology tools that enable better clinical decision support, enhanced reimbursement, and patient satisfaction to the physician community, the hospitals have in effect achieved the same loyalty outcome at a fraction of the cost. We can clinically integrate a community of 500 independent physicians for just under $1 million a year (on a software-as-a-service model). The doctors remain independent, the hospitals continue to get their admissions/ancillary services, and the patients benefit. The ROI is enormous relative to a practice acquisition strategy.

Kaiser Health News also cautions: "In effect, ACOs are an attempt to build integrated health systems like the Mayo Clinic where none exist. But Mayo took several decades to become a global destination for health care. The studies of ACOs called for in the congressional proposals aim to see if one can be formed in a year or two."

Stockard demurs:

> Even if the reform legislation twists and turns and ACOs morph into other entity forms, the fundamental concepts we are demonstrating about improving community health metrics will continue to be important.
>
> We know that with the amount and type of data that we have access to, the sky is the limit for predictive modeling, risk adjustment and population-based studies. The next evolution for us will be benchmarking relative to national norms and population insight and what doctors can anticipate relative to the population.

Chapter 12

Physical: Why Test Driving Is Still Important Even in a Digital World

You look around—the music is flowing from everywhere. Surely there must be speakers embedded in the ceiling and the walls beyond the ones under the dark drapes in this home theater. Suddenly, the host interrupts the show. Pardon me, he says as he plucks the speakers from the wall and undrapes them. They are faux logs of wood! Then he restarts the show and the music flows again. As he walks out, he promises there are no hidden speakers in the ceiling or walls.

This is no home theater. It is the Bose store at a mall. Bose as in famous for its audio technology. It is showing off its VideoWave entertainment system with an impressive screen (46-inch CCFL backlit), and its even more impressive in-built seven speaker array. It also packages its Phaseguide radiator technology that targets sounds in different areas of the room. Sure, you could watch the demo of the product at home on YouTube, but it does not come close to matching the richness of the sounds and colors in that in-store experience.

Welcome to a baffling trend: Even as Amazon and Netflix threaten all kinds of brick-and-mortar, consumers have shown they want to test-drive high-tech products like they do cars. They want to look at optional add-ons; they want coaching and service. But they don't want it to be like the high-pressure experience that comes with buying a car. In fact, Steve Jobs captured the consumer sentiment at a show in early 2001: "Buying a car is no longer the worst purchasing experience. Buying a computer is now No. 1."[1]

Test Driving Everything

Starting in March 2011, GE kicked off its Electric Vehicle (EV) Experience Tour in several cities across the United States. The goal was to "bring GE experts together with local businesses, industry leaders, and public sector stakeholders for educational workshops, test drives, and dialogue on the business case for EV ecosystems." Blogged *Engadget* after the San Francisco stop, "The Yves Behar-designed GE WattStation EV charger was on display at the event in both mock-up and ice sculpture form. We spent some time chatting with Luis Ramirez, CEO of GE Energy Industrial Solutions, and Clarence Nunn, President and CEO of GE Capital Fleet Services, about the future of EV charging. We discussed efforts like PlugShare and the recent addition to EV charging stations to Google Maps, concepts like smart parking spots with embedded inductive charging, as well as ways to accommodate folks without garages who park their vehicles on city streets."[2]

The website helpfully provides the GPS coordinates as N41° 53.655 | W87° 37.450. Very appropriate as they point to a store of the navigation company Garmin. That's 633 N. Michigan Avenue on Chicago's swanky Magnificent Mile. The website also provides a 360-degree sweeping video of the store, but you have to step in the store if you want to touch and feel the entire Garmin product line under one roof. Garmin has a very wide variety of navigation units—specialized marine ones, others for pilots of private planes, and of course plenty of devices for autos, hiking, and other uses. The store has another advantage. You can walk in and rent units for as few as three days.

Walk into the Citi Union Square branch at 14th Street and Broadway in New York City, and you will think you are in an Apple Store. That's because it was designed by the same firm, Eight Inc., that also helps design Apple Stores. The 9,700-square-foot store "features six interactive sales walls; image ATMs; free online access and wi-fi for customers; 24/7 access from the ATM lobby, to customer service experts via videoconferencing; and a private seating lounge for customers of Citigold, Citi's premium banking service."[3]

Former U.S. Vice President and Nobel Laureate Al Gore joined John Chambers, CEO of Cisco, on stage at the VoiceCon event in Orlando in 2008. The unique aspect was Gore was home in Nashville, Tennessee, and Chambers was in San Jose, California, and the discussion on climate change was moderated by a journalist in London. It was a great way to showcase Cisco's Telepresence product and Gore's *Inconvenient Truth* message with the fact "that dispersed groups can meet without flying to the same place, thus not contributing to carbon-dioxide pollution."[4] The attendees at the event did even better. They got to test-drive the experience firsthand at the massive booth Cisco set up. As Cisco described it, "(the) creative, interactive, customer experience focused booth was the 'hottest' place to be on the show floor. We created an office, small business, airport lounge, call center, and home environments for attendees to step into and experience the many solutions Cisco provides to solve business challenges."[5]

Even as the consumer bought all kinds of technology and got used to technology-enabled services, the past decade has brought an even more baffling trend. High-tech customer channels, particularly in retail, went through all kinds of turmoil.

When Category Killers Themselves Die

In his 2005 book Robert Spector called category killers "the most disruptive concept in retailing."[6] Their goal, he wrote, "is to dominate the category [e.g., toys, office supplies, home improvement] and kill the competition—whether it be mom-and-pop stores, smaller regional chains, or general merchandise stores that cannot compete on price and/or location." He said they "have helped to expand and upscale the 'mass market' by aggressively driving down the prices of goods and services."

Yet high-tech category killers like Circuit City, CompUSA, and Dixons (in the UK and Europe) have faded over the decade, and others like Best Buy had to make numerous adjustments.

Time magazine summarized Circuit City's issues: "Circuit City became complacent—a fatal mistake in the fiercely competitive and fast-evolving retail-electronics industry. The problems began a decade ago, when Circuit City failed to secure prime real estate—its out-of-the-way locations were often just inconvenient enough to tempt customers to head to other retailers, like Wal-Mart. Then Circuit City stopped selling appliances. It didn't move as aggressively into gaming as it should have. And it missed out on big in-store promotions with thriving companies like Apple Computer."[7]

A blogger captured CompUSA's issues: "I had been an infrequent customer of CompUSA—not-so-affectionately known by some of its customers as CompUSSR—over the years. I prefer to buy my hardware online, or make a 30-minute pilgrimage to Fry's when it offers something like a $169 500 GB hard drive. As I suspect is the case for many others like me, CompUSA has become the store of last resort, the place you go when you need a new router *right now*. That's not a good niche, especially when consumers are fickle and the competition is cutthroat."[8]

The *Financial Times* of London confirmed the trend across the pond, stating, "High street chains selling electrical goods may have conceded defeat in the price war with supermarkets and online retailers, but the new battleground is customer service. Advances in technology mean shoppers at Currys, Comet, or PC World stores can use their smartphones to quickly compare prices for the products on show with the same goods available online."[9]

The Best Buy Exception

An AP analysis in 2007 compared Best Buy to Circuit City: "For sure, bargains and good rebates could be found at the stores of either chain—an important draw for the price-conscious American public. At other times, it's as basic as how a store feels, how the products and aisles are laid out, how the workers there treat you."[10]

Forrester, the research firm, summarized multiple aspects of the Best Buy recipe:

> First, it leveraged its deep knowledge of consumers to understand what consumer needs were unmet in the market. Second, it approached PC OEMs to develop a unique line of laptops to meet consumers' specifications. Third, it bundled the laptops with extended warranties, antivirus protection, and/or Geek Squad service. Finally, it created its own category brand for these laptops.[11]

Start with customer research. "The company has made an art form out of customer segmentation. It talks in terms of 'Buzz' customers—young gadget enthusiasts—and 'Jill' stores aimed at suburban moms. Beyond this high-level segmentation, Best Buy has 15-plus terabytes of data on over 75 million customers: products they have bought; neighborhoods they live in from shipping addresses and rebate checks; often credit and employment histories. Its sophisticated analytics has allowed it to identify that a sliver—just 7 percent of its customers—drive 43 percent of the company's overall sales volume. So valuable to the company are these individuals that it would take approximately 39 'Uncommitted Customers' (which account for approximately 12 percent of all customers), to replace the lost lifetime value of just one of the best customers."[12]

Add Best Buy's emphasis on after-sales service with a unit called Geek Squad and projects like the Remote Service Project. "We'd launched Geek [Squad], and we wanted to maximize the availability of the 18,000 Geeks. We wanted to improve their productivity, which would improve the service to the customer and improve the time that the consumer had access to their product, so that when their PC went down, instead of losing it for a week to 10 days, we were now [shortening] downtime to 24 hours: That's a massive breakthrough. [The Remote Service Project] also meant that when you go into the store and you want your PC repaired, instead of [having a long wait time because of] that particular store having a backlog of work to do, we now have a remote capability that identifies that backlog and [can assign] the work [to another] store remotely. And so it really means that we can level out the work across the enterprise, but above all, deliver a great experience for the customer and maximize the up time of the capability of the product that they've sourced from us."[13]

Best Buy has also become a venture capitalist investing in a fund focused on digital media startup investments like games and mobile applications. An executive was quoted, "We're trying to change what the solutions are for our customers and change the direction of our company and get involved a lot more in what's happening in digital entertainment as an active participant, rather than a passive participant."[14]

Best Buy has also innovated with social media, particularly "Twelpforce" (i.e., Twitter help force), a social media experiment in customer service."[15]

Says John Burnier, Manager, Emerging Platforms and Social Media Steward at Best Buy, about Twelpforce: "With 45,000+ responses provided to incoming questions, I think we're given the opportunity to prove ourselves every day." Pressed for some examples of outstanding service, he says:

> There is one story that stands out involving one of our team members who helped a man obtain a warranty claim after a hose on a washing machine we installed went bad. He was irate, but @coral_bestbuy took him through the process, and helped turn him from a customer that may never have shopped us again, to one that will likely always shop us.

And

> I was pleased to escalate on behalf of a customer with a laptop battery which powered far less time that the specification advertised. Our store would only offer him the refund option but he liked the big screen/lightweight model and did not really want to return it. The manufacturer kept wanting to run tests. I escalated to our VP of Merchandising who called the manufacturer's rep and they moved swiftly and sent the customer a complimentary extended-life battery worth almost $200.

Yet the last couple of years have been challenging one, even for someone like Best Buy with so many unique assets. "The electronics seller was one of the few major retailers that reported an outright sales decline in December (2010), following a dismal third quarter dragged down by weak domestic revenue."[16]

To adjust, Best Buy announced it would focus "on a profitable mix of accessories, subscriptions, content delivery, and services, and will reduce its big-box real estate by 10 percent over the next three to five years, as part of an aggressive plan to build its business."[17] It is "opening 150 smaller mobile-only stores by the end of the year, nearly doubling its total to 325. The move is part of the company's bid to remake its business by focusing on more profitable, fast-growing categories such as tablet computers and smart phones."[18] Best Buy also qualified as an early retailer for the GM OnStar FMV and the Ford charging station described in Chapter 1.

In the mobile space, that means competition from an even more focused category killer—the estimated 70,000 U.S. stores of the major telcos like Verizon, AT&T, and Virgin. Overseas it would mean competing with TheMobileStore, a specialty chain in India started by its Essar Group and in China, Foxconn's (Apple's contract manufacturer) with its large retail plans. Best Buy made its business by exploiting its customer knowledge. These specialty stores are using their product knowledge to come at the market in the other direction.

Apple as the Real Category Killer

"I am not a genius, but I will stand behind this bar."

A modest Steve Jobs says that about the Genius Bar in a video sneak tour of his about-to-be-opened retail store at Tyson's Corner in Virginia. It was 2001 and people called him anything but modest or a genius for investing in brick and mortar when it looked like the whole world was going digital. Look at the video a decade later and it captures the spirit of the Apple juggernaut. It shows well-designed products, plenty of usable software, lots of accessories, and warm, welcoming, and knowledgeable service.

Over 230 million visitors entered one of Apple's over 300 stores around the world in 2010. They spent plenty of time and money there. Many also spent plenty of time outside it as crowds snake for hours with every new product introduction. Many visit the store as they would a museum. The Fifth Avenue store in New York—the glass cube—is now considered one of the most photographed sites in the city. The store at the Louvre in Paris is poetically symbiotic. It's hard to tell where the world-famous museum with the Mona Lisa ends and where the Apple store starts.

Michael Gartenberg, analyst at Gartner, invokes Nordstrom, the retailer known for its legendary customer service.

> Nordstrom is taking a page from the Apple retail playbook and rolling out a series of iPod touch-based checkout devices to their retail stores. I've called Apple the Nordstrom of technology in the past, but now it seems the student has become the master.[19]

Gartenberg continues:

> More and more I hear anecdotal stories of Apple's customer service, and how an experience went from being frustrating to heroic. These become tales at cocktail parties and dinner gatherings. The net result: The type of experiential marketing that simply can't be bought, only earned.
>
> Apple has made the technology-buying experience something that is among the best consumer retail experiences around," he concluded. "I expect this trend will only continue. Now that Nordstrom has adopted the Apple experience, how long do you think it will take others to begin to adopt such technology as well?

The Customer Channel: Own versus Outsource?

The Apple experience is making several companies think about owning their own storefronts. Google is opening Android stores with the first in Australia. Microsoft has been opening its own stores across the western part of the United States.[20] "Getting that direct customer feedback is what we're learning and getting from our stores," says Microsoft as it looks to open another 75 stores over the next few years. Nokia has opened and already closed many stores around the world, including a flagship store in London.[21]

In the meantime, medical device companies are finding it difficult to ignore Walgreen's. The pharmacy chain has over 7,500 drugstore locations and close to 6 million daily visitors to its stores.[22] Walgreen's is increasingly tech savvy, and so are its customers. So, diabetics can find supplies for their OneTouch Ping (a glucose meter that transmits data wirelessly to the insulin pump at the patient's belt level) or a Novolog, a prefilled insulin pen, at Walgreen's. Walgreen's has a voice-activated refill system. You punch in your prescription code and wait for a text message when the order is ready to pick up. Many branches have a drive-through with chutes like banks do where you can send your credit card and receive the medication. Its website makes refills even easier, since you can pick off a list of previous prescriptions. In the mobile age, Walgreen's has added Refill by Scan, which allows you to scan the bar code from a previous prescription using an iPhone or Android phone.

Companies like Toro or Stanley Black & Decker, with their own versions of smart products for homes and lawns, cannot afford to ignore a Home Depot, with over 2,000 locations around the world. Home Depot aisles are filled with many technology-rich products. They include a remote-controlled Hunter ceiling fan, a Filtrete wi-fi thermostat, a Ridgid 7,000-watt generator, a Ryobi Duet Power Paint Tool system, a Rain Bird irrigation system (to program lawn sprinklers), a Dyson vacuum with "root cyclone technology," a Char-Broil grill with infrared heat. Home Depot caters to an increasingly tech-savvy contractor and do-it-yourself customer base. Indeed, it is one of the first retailers pioneering mobile payments.[23]

Home Depot employees are themselves increasingly tech-savvy. Walk up to an employee and he or she can punch codes into Motorola handhelds and tell you the exact aisle and bin number for the item you are looking for. The Motorola also functions as a walkie-talkie, indispensable in the big stores, and as a portable cash register, and can be used to send requests for bar codes to be printed.

Technology in Retail

We have discussed the trends in high-tech retail. Just as impressive is how technology is reshaping all other types of retail.

Here are some examples:

- AT&T retail stores were some of the first to introduce Microsoft Surface computers. When a customer placed one or more phones on the table, information about features popped up. Shoppers could also zoom around AT&T's coverage map and learn about calling plans by moving their hands across the screen.[24]
- Trader Joe's has already acquired a cult-like following even as they use sophisticated analytics to only "carry 4,000 different products, compared to typical grocery stores' 50,000."[25]
- Adidas will be rolling out a futuristic Adiverse digital shopping wall at select stores. A typical shoe store for Adidas stocks around 200 shoes, well short of the 4,000 offered by the company. The virtual wall will allow the entire range to be displayed on screen while offering more information than is typically displayed in-store.[26]
- There's Zara, which thrives on fashion fads. "While other high-fashion retailers spend weeks or months waiting for low-cost suppliers scattered from China to Uruguay, Zara needs three weeks for new product development, compared to the nine-month industry average, and launches around 10,000 new designs each year. Timing is everything in the world of fashion, and to meet its timing requirements, Zara innovates by making multiple supply chains dance to the tune of high fashion."[27] Think Zara is just for us commoners? "They may have worn custom Alexander McQueen the day of the royal wedding, but the morning after, both Kate Middleton and her sister, Pippa, had returned to their girl-next-door styles. Newlywed Kate, who strolled the palace grounds with her husband, looked effortlessly chic Saturday morning in a black blazer, Greta wedges by L.K. Bennett (to be rereleased as the Kate wedge for $265 in June), and a blue pleated dress by Zara, just $90. And great minds must think alike: Pippa was spotted out in London, too, sporting crisp white pants, cute nude flats, and a bright blue tailored blazer, $100, also by Zara."[28]
- Or there is the opposite of Zara—Net-a-Porter, a London-based e-tailer of luxury-brand women's wear. During the last recession sales soared 53 percent in 2008. "Key to its success is the premium it places on service—it offers, for example, same-day deliveries in London and Manhattan. . . . Net-a-Porter's nonstop growth has helped spark a big push for online sales by the world's luxury brands—a move

many of them long resisted. The presumption was that buyers of pricey goods demanded person-to-person selling."[29]

- Of course, there are social and mobile technologies dramatically impacting retail. FourSquare lets you offer promotions for new customers and recognize your best customers. Customers share their location and the deals they've discovered there and brag about becoming "Mayor" of various sites. ShopKick rewards you if you merely walk in and open its app in stores like Target, Best Buy, Macy's, Sports Authority, and Crate & Barrel. There's ShopSavvy, which allows you to scan an item's bar code using your phone camera and that provides a list of online and local prices. There is, of course, Groupon, which is allowing retailers to bring in deal-driven business.

Physical Presence and Drama

People marvel at the logistics of launching Apple products. How do they keep products secure and secretive as they ship to so many locations? Fergus Rooney, CEO of Chicago-based agency EA, a marketing firm that helps companies with product launches, describes an even bigger launch. His firm helped Boeing plan the premiere of the next-generation 747, the 747–8 Intercontinental. He describes the challenge and the drama:

> EA concealed the plane with a massive 61-by-225-foot kabuki (a massive fabric panel that is quickly dropped to reveal something behind it) drape, which was lit with the "Incredible, Again" logo and color scheme. Multiple stages in front of the drape provided elevated spaces for the speakers and performers to address the 10,000-person crowd. Audience chairs were set up to face the stages, allowing an optimal viewing position for the reveal. Five HD projection screens played two custom-created videos highlighting the "moments" that went into the plane. The screens also projected a live feed of the program for audience members who could not see the stage up close.
>
> At the beginning of the program, lighting cast an initial bluish, pre-dawn glow on the space. Throughout the course of the event, the lighting slowly brightened to emulate a sunrise. When the kabuki drape fell to reveal the plane, its bright orange livery symbolized the sun, bringing light into the entire space.

But in preparation for the reveal day, we had to conceal the 250-foot plane's livery by covering the entire plane in brown paper and then wrapping it in plastic in Boeing's paint facility. In the middle of the night, the plane was rolled into the bay, unwrapped and draped for the reveal. Another challenge was designing/producing an event within the space of an enormous working factory. EA had to work around a limited load-in/load-out schedule, respecting the fact that the hangar was first and foremost a working factory. Airplanes had to be moved out of the manufacturing bays for the reveal then rolled back afterward to resume production. Using a kabuki drop also presented its own production challenges. The biggest of these was the massive wind-tunnel effect created by the opening of the enormous hangar doors.

Yes, physical presence can be challenging to manage, but can also yield dramatic results.

Conclusion

We would not buy a car without a test drive. Somehow we seem to have forgotten that physical, tactile experience continues to be important with technology products. The technology elite have not forgotten and they also know you need knowledgeable, friendly customer service to go with it. Let's next see how Taubman Shopping Centers have thrived in the last several decades, when brick and mortar was supposed to be dying.

Case Study: Taubman Shopping Centers

Taubman Centers, Inc. operates as a real estate investment trust (REIT). Like other real estate investors like Simon Property Group and Rodamco, it owns, manages, and leases regional retail shopping centers. Since its foundation in 1950, Taubman has developed more than 80 million square feet of retail and mixed-use properties. Taubman is distinguished by "creating extraordinary retail properties where customers choose to shop, dine, and be entertained; and where retailers can thrive"; its focus is "on dominant retail malls with the highest average sales productivity in the nation."[30] Most of its properties are in the United States, although it is growing rapidly overseas with an Asian subsidiary.

In November and December 2010, at several of its centers Taubman set up Ice Palaces and hosted an "immersive and interactive" experience. Holiday shoppers got a sneak peek into the wondrous land of Narnia, including scenes from the new movie, a light show, and snow. Actors from the movie streamed live from The Beverly Center event to 15 other Taubman shopping centers across the country, where children waved LED wands to simultaneously illuminate the Ice Palaces. In 2011, several of the centers added polar footage from *BBC Earth* to the show.

In previous holiday seasons, Taubman centers had Snow Globes and themes centered on other movies like *The Polar Express*.

Taubman has used technology around other promotions. In 2007, with a "back-to-school" promotion, many centers allowed students to create customized avatars, created with Auto Photo technology from Oddcast. In 2008, the back-to-school theme was "Yearbook Yourself" and featured a website that allowed teens to see what they might have looked like when their parents were in school, all while catching up on current back-to-school fashions. They could move from the 1950s through the 1990s and see themselves with the hair and fashion styles of the times while listening to music inspired by that era. In 2009, they enhanced the Yearbook website with integration to Facebook and other enhancements.

In 2010 Taubman announced an alliance with Sharp Electronics Corporation to equip several of its shopping centers with high-definition television lounges for shoppers. The AQUOS Entertainment Lounges feature HD televisions, with screen sizes ranging from 42 to 65 inches and content from a variety of sources, including live sports, entertainment, and news programming in each community.[31]

Do you see the pattern here? Taubman says 10 out of the first 15 Apple stores were at their centers. Now it has Apple stores in 65 percent of its malls. Beyond Apple, the centers showcase other technology stores and technology promotions and events.

Wasn't brick and mortar supposed to be dead by now? What is an old-fashioned mall doing with so much technology?

The families who went over the holidays to The Shops at Willow Bend, the Taubman mall in the northern Dallas suburb of Plano, got to feast on the 30-foot-tall color-changing Ice Palace with state-of-the-art audio and visual effects. They also got a chance to visit the Apple store in the mall. The Adrelina "extreme sports" store with a Flow Rider provided an interesting indoor surfing experience. They got to see one of the first Best Buy Mobile stores—a smaller footprint concept designed for malls.

Nick Dembla, who sells software to the mortgage industry at docVelocity and lives in the nearby town of Frisco, says, "My kids have frequented the Willow Bend mall since they were toddlers. They really look forward to the holiday Ice Palaces and other festivities. My 12-year-old daughter, in particular, has inherited my 'techieness,' so we tend to gravitate to the electronics stores. But we pale compared to one of our neighbors who is always at the Apple store there with his kids."

Charlie Bess, who works for EDS, now part of HP, in Plano and is happiest around technology, says, "It is one of those rare malls that when you go in, you don't exactly know what you're going to find. Its food court is more varied than the standard mall fare. There is also an eclectic mix of stores that draw you in, just to see what's there. It's a diverse combination of retailers that are more than the typical mall mix of anchor stores and 'the rest' that I believe appeals to a wide range of consumers."

The Mall at Short Hills in New Jersey also had a similar Ice Palace and an Apple store. In addition, music fans could walk into a Bose store and check out the latest in audio gear (and its VideoWave described earlier in this chapter). Or they could visit a Steinway store and try out much older technology in its grand piano, which still takes a year to construct. As with the Plano center, there were several more technology-themed stores.

Same with the International Plaza in Tampa, Florida. Yes to Ice Palace. Yes to Apple store. In addition, there is a Sony Style store with its laptops, cameras, and camcorders, TV and Home Entertainment. There is also a Clear store that offers 4G Internet plans for home and on-the-go.

Nina Mahoney, Marketing Director at the Tampa location, says, "We don't just target technology customers or technology tenants. Taubman shopping centers have evolved over the last decade to become destinations for both shopping and entertainment. Clearly technology is an integral part of that in today's lifestyle."

As the *Wall Street Journal* pointed out, technology, particularly Apple, is also getting a bigger share of our wallets. "In Apple's fiscal year through September, it had sales of $34.1 million per retail store. Macy's much larger stores generated $29 million on average in sales last year, and JCPenney just $16.1 million, estimates Michael Exstein of Credit Suisse."[32]

Mahoney continues:

> People actually want to touch and feel the product, which is so important for fashion and electronics! Any customer can order Apple items online, but they would rather see the new product now, experience it and feel the item, talk to the professionals and buy the hot product now.
>
> People come with friends and family to enjoy our many experiential stores—Disney's newest prototype store, Apple, Bose, Clear, Destination Maternity, Fit 2 Run, Lush, Sephora, and Sony Style, along with a wide choice of 16 full-service restaurants.

One of those restaurants has been fairly influential in the local technology community. Fritz Eichelberger, founder of Hotspaces.Net, a consulting and recruiting firm, explains: "I've been hosting the 'Pure & Shameless' Tech Socials at the Blue Martini (tapas bar and cocktail lounge) for almost a decade. The location is a favorite among attendees. It is conveniently located in the Tampa Bay area, there is plenty of parking, and there are other activities at the International Plaza to entice my attendees to show up. I usually get 200 to 250 attendees and once had 400+."

Thomas Neudenberger is Chief Operating Officer of the U.S. subsidiary of realtime AG, a 25-year-old German SAP consulting company. It offers a biometric SAP security product called bioLock. Looking for a city to locate U.S. operations in 2002, he evaluated several East Coast cities. While he liked Tampa, he was unsure of the local technical talent. The manager of the World Trade Center office space he was evaluating in Tampa took him to one of Eichelberger's events.

Says Neudenberger, "The Blue Martini made my decision as the quality of local talent impressed me. I moved my family here two weeks later. I have been going to that event ever since and meet many interesting people with whom to discuss our bioLock technology. It is always fun and you always run into people that you don't expect."

Nine years later Neudenberger says he made the right decision and his unit has expanded even outside the United States and installed bioLock software in Europe, Africa, and South America.

Mahoney continues:

> Many customers tell us they feel like they are on vacation while shopping/dining at IP (shoppers call us IP)! And many of them are—they stay at the Marriott Renaissance hotel attached to the mall.
>
> IP has a lobby lounge appeal, with beautiful furniture throughout for shoppers to relax and catch up on emails if necessary. We've had mall-wide/parking lot wide wi-fi for over five years (free to Brighthouse Network—the local cable provider—customers); Starbucks and Yogurbella also have wi-fi hot spots.
>
> We also offer a highly interactive children's play area that is also very educational. Families will travel past other shopping centers to bring their children to our play area.
>
> We were the first shopping center company (10 years and running) to create and sustain a weekly e-bulletin (e-blast) service for shoppers who could customize which store offers they received.
>
> So, in many ways we have invested heavily in creating a digital infrastructure.

There is also a growing trend to use the mall as a showplace to launch technology products. Mahoney again: "As you can see with the Sharp Aquos lounge, we provide new products plenty of exposure. We had the

Nissan Leaf (the electric car) exhibit here. We have had Sony Reader events here."

Technology was supposed to kill brick and mortar. Taubman is proof that what doesn't kill you makes you stronger!

Chapter 13

Paranoid: But Not Paralyzed

As companies develop their smart products, they need to become thick-skinned about what are generally accepted traditions in the tech industry.

A good example is teardowns as soon as products become available. Photos of components, guesses about their likely suppliers, likely costs, and product margins show up on blogs, the most prominent of which is a website called iFixit.

"An iFixit teardown is at once a twenty-first-century repair manual, a work of art, an exhibition of a curiosity, and an activist gesture."[1]

Some of the advice and related parts available on the site are for the aficionado. Example: "Apple has substituted (on the Verizon iPhone 4) the two bottom Phillips screws near the dock connector with 5-point 'Pentalobular' screws. This guide will illustrate how to replace the Pentalobe screws with good ol' regular Phillips screws." It has for sale a "liberation kit," which includes a 5-point "Pentalobe" screwdriver, a #00 Phillips screwdriver, and two 3.6 mm "Liberation" Phillips screws.[2]

Most users find the site even more useful for a whole variety of repair guides for cars, cameras, and other devices.

Then there are the bragging rights from the teardowns. Says *Popular Science* magazine: "In late 2006, PopSci.com had one of its all-time highest traffic days when we posted photos looking inside the then just-to-be-released Nintendo Wii. At that point, we had beaten even iFixit to the punch."

Jailbreaks and Roots

The term "jailbreak" has become synonymous with the Apple iPhone. It's in response to very high international roaming rates the carriers expect and because many users believe they are being double-charged for tethering when they are already paying for data plans.

> The first iPhone jailbreak procedure was in fact quite tedious, requiring users to download various software tools and manipulate code within the iPhone's framework. However, over time iPhone jailbreak has evolved and become a reasonably simple process. Today, all that is required to jailbreak your iPhone in most cases is to download one software tool and literally click a button.[3]

In response, Apple has periodically updated the iPhone operating system. AT&T and other carriers have implemented detection technology to identify users who have "jail-broken" their iPhones. It's a cat-and-mouse game.

The concept of jailbreak on the iPhone has evolved into what's called rooting on Android devices. "Rooting is the process by which you regain administrative access to your phone. Even though Android is an open source operating system, you still don't have full 'root access' to do what you please. Back when the iPhone launched in 2007 the hardcore techies quickly realized the true potential of the device, and the cruel software limitations that Apple had sealed it with. What became 'Jailbreaking' on the iPhone was quickly translated to other platforms as well, and when the world saw the first Android back in 2008, the term 'Rooting' was born," says the Android Authority blog.[4]

The Barnes & Noble Nook is an Android device. "The device's color touchscreen and assortment of Internet-enabled applications help differentiate it from Amazon's increasingly ubiquitous Kindle."

"Barnes & Noble intends to eventually expose more of the Nook's Android functionality to end users in future updates, but Android enthusiasts have already gotten a head start." They rooted it to make it an incredibly inexpensive tablet at its $250 (and declining) price.

The teardowns and the rootings pale in comparison to the much more malicious hacking that targets digital products and websites.

"Tarred and Feathered"

The repeated and public humiliation of Sony by hackers over a matter of months shocked the technology world. Its PS3 game console was once considered invulnerable. "But in December 2010 at the Chaos Communication Conference in Berlin a group of European programmers calling themselves fail0verflow revealed they had finally broken specific lower levels of the PS3's encryption system that let them run their own programs on the console."[5]

> Not long after that, in/famous hacker George "Geohot" Hotz decided to open up a veritable Pandora's Box of problems for Sony by rereleasing the PS3's master security key to the public. This move essentially allows anyone to run custom code on the PS3 and worse, with a bit of additional fiddling, pirated games.[6]

Blogged the Daily Tech:

> Much as Sony has abused its corporate power over users, hackers—most notably Lebanese-based Idahc (Twitter) and the international group "LulzSec" (Lulz Security)—have lorded their superior security skills over the clueless giant, constantly mocking and lashing it.[7]

Sony a clueless giant? If so, there may be many more examples.

Time magazine ran an article where it talked about hacks at Citigroup, the FBI, the CIA, Mastercard, Visa, and elsewhere.

> So who would you like to hack today? A bank, a website, a corporation or perhaps a government agency that's rubbing you the wrong way? The hacktivist group LulzSec is taking requests.[8]

If a mainstream magazine like *Time* can talk about it, think of the discussions at Black Hat conferences, which "are a series of highly technical information security conferences that bring together thought leaders from all facets of the infosec (information security) world—from the corporate and government sectors to academic and even underground researchers."[9]

Actually, the infosec world is very well aware of what are called Common Weaknesses. MITRE maintains the CWE (Common Weakness Enumeration) website, with the support of the U.S. Department of Homeland Security's National Cyber Security Division, presenting detailed descriptions of the top 25 programming errors, along with authoritative guidance for mitigating and avoiding them."[10]

The errors make for exotic language such as "Improper Neutralization of Special Elements used in an SQL Command" ("SQL Injection") and "Buffer Copy without Checking Size of Input" ("Classic Buffer Overflow").

The big problem, of course, is that the CWE site catalogs more than 800 programming, design, and architecture errors that can lead to exploitable vulnerabilities. Hackers, the white-hat and the malicious kinds, continue to add to that already long list of 800.

British tabloid *News of the World* was closed down in July 2011 over a phone-hacking scandal. The newspaper owned by Rupert Murdoch allegedly snooped on voice-mail messages of murder victims, as well as celebrities, politicians, and the British royal family. These were shady reporters. Now extrapolate that to imagine what more technology-savvy hackers can do.

Next Generation of Even More Terrifying Hacking

Richard Perkins and Mike Tassey were told that an in-flight hacking platform was impossible. In response, the pair showed off their wi-fi hacking, phone-snooping, homemade UAV at the Black Hat conference in Las Vegas in August 2011. They call their creation the Wireless Aerial Surveillance Platform; it is described in detail in our case study later in this chapter.

Researchers have demonstrated that certain pacemakers that use a wireless signal for easy tweaking are vulnerable to anyone with the correct reprogramming hardware. Doctors use these wireless programming devices to make subtle adjustments to the heart helpers without the need for further surgeries. Unfortunately, the signal they use is unencrypted, meaning that anyone who finds a way to obtain such a device could literally manipulate the heart of a patient, causing cardiac arrest or even death. [11]

At the same Black Hat conference, Don Bailey and Mathew Solnik presented how they had "found a way to unlock cars that use remote control and telemetry systems like BMW Assist, GM OnStar, Ford Sync, and Hyundai Blue Link. These systems communicate with the automaker's remote servers via standard mobile networks like GSM and CDMA—and with a clever bit of reverse engineering, the hackers were able to pose as these servers and communicate directly with a car's onboard computer via "war texting"—a riff on "war driving," the act of finding open wireless networks."[12]

Also at that conference, a pair of researchers demonstrated how home automation systems can be vulnerable to attacks. "Carrying out their research independently, [Kennedy] and [Rob Simon] came to the same conclusion—that manufacturers of this immature technology have barely spent any time or resources properly securing their wares."[13]

The vulnerability of smart cars, homes, and medical devices means we all have to plan for far smarter security.

Technology's "Area 51"

After Apple had its well-publicized antenna issues with the iPhone 4, it took a handful of journalists on a tour of its wireless testing lab.

"Apple's wireless lab has 16 different anechoic chambers—think of them as bank vaults, padded with foam shaped into pointy cones to stop all reflections, designed to create completely radio-neutral environments. Each of these chambers is estimated to have cost $1.2 million. The existence of this lab used to be secret," an Apple PR representative pointed out. "Now it's not."[14]

Google's data centers have long been even more secretive. In April 2011 Adam Swindler of the Google Enterprise group blogged, "Many of you have been interested in visiting our data centers to see how we work to protect your data, but access to them is tightly restricted. Since we can't give everyone a tour, we look for other ways to provide some visibility into these buildings. Last year we published the Google Apps security white paper, earlier this year we hosted a security and privacy webcast, and today we're sharing a video that highlights some of the capabilities in our data centers."[15]

There was widespread speculation about why Google was showcasing that video. Some thought it was in response to a recent data center outage at Amazon and they wanted to assure customers their data centers were secure. Others speculated it was in response to the more open stance Facebook had taken with its Open Compute Project (see Chapter 8), which highlighted detailed information of its recently opened Prineville data center.

If Google is secretive about its data centers, was it also helping Apple be secretive about its iCloud data center? *PC Magazine* reported:

> . . . it's a Google Earth shot of Apple's giant data center in the boonies of North Carolina. The image was previously unavailable on Google Earth, but as Apple's iCloud announcement nears, the company seems to have allowed Google to show the massive, 500,000-square-foot facility to the world.[16]

Labs and data centers are part of the industry's "Area 51," the military base in Nevada of UFO rumor fame. They exist behind prying eyes and there is usually plenty of disinformation about them. There is always risk of hostile activity. As we saw in Chapter 4 the whole country of Estonia was brought to its knees for weeks in a cyber attack widely speculated to have originated in Russia.

On its official blog Google described its investigation of a similar attack in 2009:

> Like many other well-known organizations, we face cyber attacks of varying degrees on a regular basis. In mid-December, we detected a

> highly sophisticated and targeted attack on our corporate infrastructure originating from China that resulted in the theft of intellectual property from Google. However, it soon became clear that what at first appeared to be solely a security incident—albeit a significant one—was something quite different.
>
> First, this attack was not just on Google. As part of our investigation we have discovered that at least 20 other large companies from a wide range of businesses—including the Internet, finance, technology, media, and chemical sectors—have been similarly targeted. We are currently in the process of notifying those companies, and we are also working with the relevant U.S. authorities.[17]

In 2008, as he was getting ready to unveil his 400,000-square-foot "SuperNAP' in Las Vegas, Switch CEO Rob Roy remarked he expected it to be filled by the world's most prominent companies. He was promising 100 percent, not 99.99 percent, uptime—the Holy Grail in enterprise computing.[18]

That, of course, demands extreme security measures.

The *Wall Street Journal* described some of the security at the facility:

> The guards . . . are not your typical rent-a-cops. These are Switch employees recruited from the Marines and other military services—buff, dark-uniformed hunks who sport sidearms inside the building and automatic weapons outside. They never smile.[19]

Switch could be guarding against threats from hostile countries, but even more against industrial espionage.

Dumpster Diving in the Digital Age

Gizmodo, which ended up paying for and dissecting a prototype of the iPhone 4 that an Apple employee misplaced at a bar, described its impression of Apple's security:

> At their Cupertino campus, any gadget or computer that is worth protecting is behind armored doors, with security locks with codes that change every few minutes. Prototypes are bolted to desks. Hidden in these labs, hardware, software and industrial-design elves toil separately on the same devices, without really having the complete picture of the final product.[20]

Gizmodo went on to suggest Apple's tight security was tied to marketing advantage: "The Gran Jefe Steve trusts them to avoid Apple's worst nightmare: The leak of a strategic product that could cost them millions of dollars in free marketing promotion. One that would make them lose *control* of the product news cycle."

Actually, Apple and other technology companies have the tight security to protect against many different ways trade secrets are compromised. A paper by Mark L. Krotoski of the U.S. Justice Department[21] highlights some of the scenarios based on prior cases and investigations:

- A trusted employee with access to valuable company information who, after becoming disgruntled, downloads and transmits the information to others outside the company, who offer it to the "highest bidder."
- A competitor who devises a scheme to gain access to company information for use in fulfilling an international contract.
- Employees who execute a plan to steal proprietary information and take it to another country and are stopped at the airport.
- After being offered a senior position with a direct competitor, and before tendering his resignation, an employee uses his supervisory position to request and obtain proprietary information he would not normally be entitled to access. After taking as much proprietary information as he can, he submits his resignation and takes the materials of his former employer to his new position and employer.

Of course, to go with the dark arts, there is always the dark humor. The Apple iPhone bar incident described earlier led to a series of industry jokes that go like this:

> A Microsoft/Nokia/RIM (take your pick) employee lost a secret prototype of the company's next-generation device. The person who found the prototype tried to sell it to the highest bidder and was offered just $10. Asked why, the publisher of a tech blog said, "Even $10 is too much to bring in an extra 50 page views, particularly since 40 of those will come from the vendor CEO's desk.

Ouch.

Conclusion

Teardowns, jailbreaks, and rootings—they are almost badges of honor in the technology world. They are tame, however, compared to the malicious hacking and espionage technology companies are increasingly facing. In such a climate, it helps to be paranoid. The next section on the Wireless Aerial Surveillance Platform shows other reasons to be vigilant. Of course, being paranoid does not mean being paralyzed. The technology elite just look at it as a cost of doing business. Life has to go on.

Case Study: Wireless Aerial Surveillance Platform

In late July 2011, Mike Tassey and Rich Perkins drove more than 1,700 miles from their homes in the Midwest to Las Vegas, Nevada. The drive was mostly uneventful. There were hundreds of miles of flat farmland at the front end and hundreds of miles of desert at the other end. The big excitement was a crack in the car's windshield, thanks to a mud-covered pickup truck dropping baseball size chunks of mud, rocks, and gunk onto the highway in front of them.

They were carrying some yellow cargo that could have brought law enforcement scrutiny. No, there was nothing explosive. If the law had run a background check, it would have soon found both Tassey and Perkins had fairly high-level security clearances. Both have worked for the U.S. Air Force and other federal agencies.

The cargo was definitely explosive, but in a different sense. Tassey and Perkins were carrying an FMQ-117B U.S. Army target drone to the Black Hat/DefCon conference in Las Vegas.

The 14-pound, 67-inch wingspan drone is by itself dated technology. It was first shipped in 1979 for surface-to-air defense training. The 2 × 6 cell 22.2v 5000 mAh LiPo batteries allow it to fly for less than an hour at a time, compared to the drones that go across the world in today's battles.

It was the payload and capabilities that impressed the tough-to-impress security audience the conference attracts each year. (The audience is by definition paranoid. Many pay cash for the conference and register under assumed names.)

The payload (a Via Epia 10000EG Pico ITX motherboard, 1 GHz Via C7 CPU, 1 GB of memory, storage in the form of a 32 GB Voyager GTR Flash drive, networking via a USB 4G dongle, an HD camera, and the Linux operating system BackTrack) is designed for hacking, assisted by a custom-built 340-million-word dictionary for brute-forcing passwords.

The drone is equipped with a pair of 900 MHz XBee radios that provide telemetry and data link. These channels are protected by 128-bit

AES encryption and allow the controllers on the ground to access the payload and flight computers remotely as if they were present onboard.

Just what does it do with all that technology?

In an early concept it was designed for wi-fi penetration in a target area over which it is flying. The thermal sensors on the aircraft's skin, the onboard GPS sensors to detect the horizon, and the airframe's attitude and position in space allow for an accuracy of under three meters. Navigation coordinates are preplanned using software integrated with the Google Maps repository and are as simple as clicking on a map.

In its current iteration it adds a Universal Software Radio Peripheral (USRP) made by Ettus Research, and it can intercept calls made on GSM phones (like the iPhone that uses AT&T's GSM network) into connecting with it as if it were a standard cell-phone tower. It then records any phone conversations or text messages while connecting the call via VOIP, giving the impression the call went through normally.

Says Tassey:

> What we call our Wireless Aerial Surveillance Platform is a proof-of-concept UAV (Unmanned Aerial Vehicle) designed to demonstrate the ability of a relative layman to utilize off-the-shelf and open-source components to craft an autonomous platform from which to launch attacks against wireless clients, networks and cellular phones on the ground.
>
> We made several modifications, but did not use any custom manufactured parts. As an example, we modified the airframe to utilize the electric motor to quiet the UAV so that we could operate nearly silently. We added an off-airframe processing capability that can reside anywhere on the Internet. This capability makes use of the Compute Unified Device Architecture (CUDA), which allows the use of NVidia Graphics Processing Units (GPUs) present on inexpensive video cards to process mathematical data at incredible speeds.
>
> Most of the hardware and software is easily available online. The average enthusiast can build and operate this. The whole package cost us only around $6,000 to put together.

Says Perkins:

> The thrust of the concept was that organizations spend large amounts of money on physical security (locks, doors, fences, guards, etc.), which

is focused on the "bad guy" being a person with a backpack, a car in the parking lot, or an imposter with a laptop. The more we looked at the state of security the more it became obvious that no one is looking up at the skies.

No one is looking up because, until recently, the technology and the skills needed to create a viable unmanned cyber-attack drone were out of the reach of anyone not affiliated with the government or part of a specialized group of researchers at a university. The average person thinks of the General Atomics MQ-1 Predator used by the U.S. Air Force when they hear the word drone. Those cost millions of dollars each.

That has changed. Today nearly anyone with just a few hundred dollars can design and build a powerful autonomous UAV with the capability to fly hundreds if not thousands of miles completely without intervention.

We saw that the use of UAVs to launch cyber-attacks was a threat vector that had not been considered or modeled by organizations today, but one that was very plausible given the state of technology and the availability of off-the-shelf parts. Our goal was to utilize what was available to create a drone which would illustrate the threat, forcing organizations to take a hard look at how they view information security.

Tassey continues:

The UAV allows an attacker to get inside the perimeter defenses, to spoof access points within your network and isolate clients for attack, attack your wireless infrastructure by breaking encryption and then leverage any of hundreds of onboard tools to attack internal machines, gather data, gain access to critical systems and cause service interruptions or move sensitive data out of the network through the aircraft. It really is able to get in, loot and pillage, and then get out without traversing the boundary defenses that organizations spend so much time and money building. In addition, it also gives the attacker the ability to attack GSM cellular phones, giving them the ability to monitor phone conversations and SMS messages, reroute phone numbers, and inject audio and data. Think of the volume of company proprietary data or personal information that is communicated every day via cellular phones and one begins to understand the type of damage that can be done.

The reaction at the conference was very positive. The online response to them in various communities they are part of has been mixed.

Perkins explains:

> We have received comments ranging from "I'm going to report you to the FAA/FCC [the U.S. agencies that oversee aviation and telecommunications]!" to the more violent "I'm going to shoot the drone down with a shotgun." What we found is that while the Internet has provided a medium through which like-minded people can gather and develop their ideas, there is still distrust and animosity toward outsiders or those who these communities perceive as portraying their hobby or interest in a potentially negative way. We found that many of the comments were based on fear.
>
> The gratifying thing is the positive uses people are finding for the device. Small cheap UAVs can be developed for law enforcement and the military to enhance their mission capabilities by providing instant oversight of a battleground, crime scene, or search area. Customs and Border Patrol agents on patrol could use a small hand-launched UAV equipped with low-light near-infrared cameras to locate "coyotes," smugglers, and illegals, and provide an ability to cover large areas of intractable terrain from the air. Intelligence agencies could use a small UAV to attack local target networks, gather signals intelligence, and tracking of subjects by video and electronic emissions. Search and rescue and law enforcement could use these types of UAVs to provide communications relay, drop beacons, and provide extended search capabilities without the cost and limitations of a manned aerial platform.

Tassey says over the course of the project that "We have found that there are quite a few groups doing UAV research for imagery and radio projects from around the world."

Whether their device is used positively or negatively, they deserve credit for raising the level of paranoia around tech security.

Summarizes Perkins:

"We want people to stop thinking that they can be complacent and make assumptions about security, because the bad guys aren't complacent at all. They are creative, intelligent, and always take the path of least resistance. It is truly a case of 'If we can do it . . . so can they.'"

Chapter 14

Pragmatic: When Attorneys Influence Technology Even More than Engineers

Early in 2011, the IBM supercomputer Watson thrilled the world by beating two of the most successful *Jeopardy!* contestants of all time. While IBM basked in the glory, Yahoo! reminded the world that it also deserved a fair amount of credit:

> IBM's Watson depends on 200 million pages of content and 500 gigabytes of preprocessed information to answer the *Jeopardy!* questions. That huge catalog of documents had to be indexed so that Watson could answer questions within the 3-second time limit. On a single computer, generating that large catalog and index would take a lot of time, but dividing the work onto many computers makes it much

> faster. Apache Hadoop is the industry standard framework for processing large amounts of data on many computers in parallel. By using Hadoop MapReduce, Watson's development team was able to easily and reliably run their application on a large number of computers. For the last 5 years, since the start of Hadoop, Yahoo! has been the primary contributor.[1]

A comment to that blog sought to minimize that claim: "Watson does not use Hadoop during runtime (e.g., when answering questions)—only to prepare it's (sic) corpus of source materials (text, encyclopedias, etc.)."

Another blog boasted about Yahoo!'s contribution: "Yahoo! created Hadoop and since then has been the most active contributor to Apache Hadoop, contributing over 70 percent of the code and running the world's largest Hadoop implementation, with more than 40,000 servers."[2]

Beyond Yahoo! and the Cloudera distribution of Hadoop described in Chapter 11, many more technology vendors can claim credit for various implementations of Hadoop:

> Everyone from EMC and IBM to database startups like Hadapt and DataStax (are getting) into the business of selling Hadoop-based technologies and services.[3]

You may have heard of fuzzy logic. Welcome to the world of fuzzy ownership of intellectual property. The Microsoft Kinect legal story is just as interesting.

The Kinect Legal Scenario

Soon after Microsoft introduced its very successful Kinect in 2010, its gesture-based interface to the Xbox, Adafruit Industries announced OpenKinect, a contest with a $3,000 bounty. Adafruit is a New York City–based company that sells kits and parts for original, open-source hardware electronics projects. It called for "Open-source drivers for this cool USB device, the drivers, and/or application can run on any operating system—but completely documented and under an open-source license."[4] The goal was to take the interface beyond gaming to education, robotics, and other applications.

Microsoft did not take kindly to that:

> "Microsoft does not condone the modification of its products," a company spokesperson said. "With Kinect, Microsoft built in numerous hardware and software safeguards designed to reduce the chances of product tampering. Microsoft will continue to make advances in these types of safeguards and work closely with law enforcement and product safety groups to keep Kinect tamper-resistant."[5]

Problem is, Kinect was technically not a Microsoft-developed product. It had licensed technology from PrimeSense, an Israeli company that "developed the method and core software that Microsoft's forthcoming camera/controller uses to read human bodies and gestures."[6]

Microsoft quickly backed down: "OpenKinect was not a hack, but the new code is not supported and people should use the Kinect as intended, with the Xbox only."

There are two reasons for the possible back-down, speculated Dana Blakenhorn at ZDNet:

> Microsoft lawyers recognized it has no legal case against Martin (who developed the drivers for the Adafruit bounty above), who made no changes to the hardware.
>
> Microsoft marketers realized that the drivers might, in the end, be a gold mine for Microsoft.[7]

In the meantime, PrimeSense, along with other organizations, launched the OpenNI organization to certify and promote the compatibility and interoperability of Natural Interaction devices, applications, and middleware.[8]

These IBM and Microsoft episodes are actually pretty civil compared to the all-out legal wars that have spread over technology.

Full Employment for Technology Lawyers?

It is tough to remember a time in the past when there has been so much technology-related legal activity. Glancing at the 2010 Apple

10-K and other legal filings, you see disputes involving Motorola Mobility, Inc., Nokia Corporation, Samsung, Psystar, AT&T Mobility, and many others. In 2005, the UK publication *The Inquirer* analyzed Microsoft's legal track record:

> The surprising thing is not only the number of those lawsuits against Microsoft—at one time, it had more than 130 pending—but more importantly, the sheer amount of money it represents. The Redmond giant has been ordered to pay nearly $9 billion, a figure that is set to rise with some lawsuits still to be ruled on.[9]

The EU would appear to have proceedings against every major U.S. technology company.

In turn, just about everyone seems to have litigation pending against Google over its Android mobile platform.

"The irony's enough to make your head explode: Microsoft makes more money from Android than it does from Windows Phone. This according to Citi analyst Walter Pritchard, who says $5 from the purchase price of every HTC Android handset sold ends up in Microsoft's pockets."[10] And Microsoft is rumored to want even more from other Android device makers, such as Barnes & Noble.

Oracle has sued Google directly (not the device makers) and "claims it should be awarded $2.6 billion for Google's alleged infringement of its intellectual property with regard to the Android mobile operating system."[11]

With a consortium of companies (Apple, EMC, Ericsson, Microsoft, RIM, and Sony) outbidding Google for a portfolio of Nortel's wireless patent portfolio, it could mean more legal hassles for Android. "Google had been expected to emerge victorious after it set a $900 million stalking horse bid in April. But the auction that started on Monday and saw 20 rounds of bids over four long days ultimately hit a price that became too much even for Google, Reuters reported."[12]

Google, in turn, looked to the government for relief, "calling on Congress and the Federal Trade Commission to rein in lawsuits, and asking the U.S. Patent and Trademark Office to take closer looks at patents being used in litigation."[13]

Google also turned to buying pending and issued patents from IBM Corp. in its quest to shore up its defenses against suits by other technology companies, according to documents filed with the U.S. Patent and Trademark Office.[14]

Google's biggest move came in the form of its $12.5 billion bid for Motorola Mobility. While it gets plenty of engineers and mobile products from the acquisition, the biggest asset it acquired was Motorola's large vault of approved patents and in-progress patent applications. Larry Page, CEO of Google, did not mince words on the Google blog:

> We recently explained how companies including Microsoft and Apple are banding together in anti-competitive patent attacks on Android. The U.S. Department of Justice had to intervene in the results of one recent patent auction to "protect competition and innovation in the open source software community" and it is currently looking into the results of the Nortel auction. Our acquisition of Motorola will increase competition by strengthening Google's patent portfolio, which will enable us to better protect Android from anti-competitive threats from Microsoft, Apple, and other companies.[15]

Verizon's 2010 10-K says it and a number of other telecommunications companies:

> have been the subject of multiple class action suits concerning its alleged participation in intelligence-gathering activities allegedly carried out by the federal government, at the direction of the President of the United States, as part of the government's post-September 11 program to prevent terrorist attacks. Plaintiffs generally allege that Verizon has participated by permitting the government to gain access to the content of its subscribers' telephone calls and/or records concerning those calls and that such action violates federal and/or state constitutional and statutory law.[16]

Verizon's 2009 10-K showed even more diverse disputes:

> Our business faces a substantial amount of litigation, including, from time to time, patent infringement lawsuits, antitrust class actions, wage and hour class actions, personal injury claims, and lawsuits relating to our advertising, sales, billing, and collection practices. In addition, our wireless business also faces personal injury and consumer class action lawsuits relating to alleged health effects of wireless phones or radio frequency transmitters, and class action lawsuits that challenge

> marketing practices and disclosures relating to alleged adverse health effects of handheld wireless phones. We may incur significant expenses in defending these lawsuits. In addition, we may be required to pay significant cash awards or settlements.

Let's Go to Marshall, Texas

Many technology lawsuits end up in Marshall, Texas, because "in the rough calculus of intellectual property litigation, [its] tough judges equate with speedy cases—and that's exactly what you want if you're a plaintiff with limited cash, but potentially big-time settlement payments or damages from a company you claim is infringing on your patent."[17]

So the town of 20,000 "with more pottery manufacturers than software companies" has become famous around the world. It comes complete with its folklore of a "rocket docket" for the speed of its cases and of lawyers making "rattlesnake speeches" similar to the loud posturing the local venomous species does to warn of its presence.

Willie and Waylon and the boys put Luckenbach, Texas, on the map. Technology attorneys have done even better for Marshall.

Why This Level of Heated Legal Activity?

Several major trends are converging to cause this heightened level of legal activity:

- The high level of acquisition activity in legacy vendors over the past decade and their sense of entitlement that they need to squeeze every last drop from the assets they acquired.
- Several new mobile, social, and other markets where patents seem to infringe and cross-licensing agreements have not yet stabilized.
- A bit of the jealousy factor about the "Johnny come latelies." Let's face it—Apple was not a player in mobile markets a few years ago. Google was not a major analytics player. Amazon was not an outsourcing vendor. It is tempting for incumbents to challenge them in court even if they cannot in the marketplace.

- Growing corporate discontent with poor payback and quality for the billions they have spent on IT. It is a miracle we do not have hundreds more lawsuits around ERP projects and data center outsourcing performance and pricing.
- Consumer frustration with poor mobile and other broadband service, combined with outrageous and creative roaming, early termination, overage, underage, and other charges. Then there is their discontent when it comes to warranty service around most technology products, virus protection, and other areas.
- Privacy issues that have grown as mobile, social, location-based services, and smart devices share more and more of our personal data, and companies do not show enough responsibility in protecting that data or restraint in profiting from it.
- Growing pressure on regulators to defend consumer rights when it comes to privacy, net neutrality, and price/performance benchmarks.

Time for a Technology Apalachin?

We have seen it in the movie *The Godfather*. We have also seen it hilariously parodied by Billy Crystal in the movie *Analyze This*. It is a meeting where the Cosa Nostra—the major Mafia families—come together, especially during times where rivalries seem to be getting out of hand. One of the few publicly known meetings was at Apalachin, NY, in November 1957, at the home of mobster Joseph "Joe the Barber" Barbara. The sheer number of expensive cars that showed up raised eyebrows of local authorities. In a comic twist, during the raid, many of the crime bosses fled into the woods. There were rumors of large currency notes being found months after the incident as the bosses dumped incriminating evidence. More than 60 underworld bosses were detained and indicted. The incident supposedly also got the FBI to wake up to the fact that the Mafia was a reality.[18]

It may be time for a Technology Apalachin. The industry leadership and regulators need to agree to tone down the legal noise and agree that all this litigation expense should be going toward more R&D and innovation.

That would mean tackling, among other things, the minefield of software patents. Brad Feld, a Managing Director at Foundry Group, has an enviable track record as an early-stage investor and entrepreneur for over 20 years. He says:

> For a while, I felt like I was shouting alone in the wilderness. While a bunch of software engineers I know thought software patents were bogus, I had trouble getting anyone else to speak out against software patents. In the last few months, however, the issue of software patents—and the fundamental issues with them—have started to be front and center in the discussion about innovation. NPR has run a couple of terrific shows on the topic. Google has obviously been making noises about it given all the legal and patent acquisition activities they are involved in. But my favorite was a blog post by Mark Cuban, which summarized, "We need to face the facts, patent law is killing job creation. If the current administration wants to improve job creation, change patent law and watch jobs among small technology companies develop instantly."
>
> I want to reiterate what I have been saying for a while, to all the software engineers who are co-authors on patents that they aren't proud of, think are bogus, were forced to create the patent by their company, were paid a bonus by their company to write a patent on nothing, are now working for a company that is getting sued for a patent they co-authored that they aren't even sure what it says. Speak Up!

But instead of hoping and praying for such an Apalachin event to take place, it helps to think through early in its lifecycle the wide range of legal issues around a technology-embedded product. Benjamin Kern, an attorney at the law firm of McGuire Woods, provides a long list of considerations in the next section.

Conclusion

If the hacking and the espionage described in the previous chapter does not paralyze companies looking to build smart products, the spreading lawsuits surely can. In many ways, the technology elite know that good lawyers are just as important as engineers in technology. They can help enterprises to be pragmatic, not paralyzed, even when surrounded by "rattlesnakes." Benjamin Kern, an attorney, provides some of that pragmatic advice in the next section.

Guest Column: Legal Considerations in Technology Product Launches—Benjamin Kern

Benjamin Kern is a partner in the technology and outsourcing practice at McGuireWoods, an AmLaw 100 firm, with approximately 900 lawyers and 17 strategically located offices worldwide. His practice focuses exclusively on transactional business matters driven by technology, including licensing, services, outsourcing, development, and technology commercialization, as well as venture capital funding, mergers, and acquisitions.

Much of my practice since the late 1990s has focused on contracts between innovators and the companies that help them finance, design, build, manufacture, market, sell, or license their innovations. In the case of new devices, a contract often starts with a framework that details the responsibility and contributions of each party. For example, the inventor of a gadget may agree to license a patent or know-how to a larger company that has the ability to bring that gadget to market. Or two companies with complementary strengths or technologies may team up to make a device that combines their capabilities. In each of these cases, there are many costs, responsibilities, and liability exposures that must be considered and allocated between the parties.

Working with these companies and these contracts proved instructive when I launched my own gadget. Canary Wireless was a virtual company that designed, manufactured, and sold the world's first LCD wi-fi detection device. Canary launched its first product in the 2004 holiday season, with no physical office, a design partner in the Chicago suburbs, an outsourced shipping company that held inventory in Green Bay, Wisconsin, and a manufacturer in Taiwan. By early 2006, within barely a year of making its product available, the company had reached seven figures in revenue. With a Yahoo!-based e-commerce site, a virtual presence, and no prior relationships with customers or resellers, the company had filled website orders to customers in all 50 states and more than 30 countries.

In running my own product initiative, and in advising big and small companies who have launched products on a regular basis as the global marketplace has increasingly embraced connectivity, I have come across many pitfalls and legal hazards that must be navigated by product companies. Here, I outline some of the prominent ones.

1. Patents or Trade Secrets

It may be helpful to think of a patent review in two parts: (1) a freedom to operate analysis, and (2) an analysis of the ability to obtain protection for a new technology. The freedom to operate analysis focuses on published or issued patents, and in particular on any claims that may cover a portion of the hardware or software that comprise the device. In cases where multiple existing patent claims may overlap with a product, inventors may prepare a matrix to assist in determining where problem areas are, where changes or other workarounds may be possible, and where licensing may be necessary. This review may also need to encompass overseas or international patent and intellectual property laws. Companies that hire contractors or outsourcers to provide design or development services may want to ask for indemnification, which can provide some financial protection if the design created by a contractor results in infringement of an existing patent.

The analysis of whether an invention can be protected by patent registration may be a separate process from the freedom to operate analysis. In some cases, as where an invention is covered by expired patent claims, an invention may not raise infringement concerns, but also may not lend itself to protection under the patent laws. Alternatively, an inventor may have missed a patent deadline (this is a more common issue in light of recent changes to U.S. patent law), may be particularly sensitive about the publication of certain information regarding a device's processes, or may otherwise elect to seek protection of a device or an invention as a trade secret, instead of going through the patent protection process. When a new product is invented, the inventor or his or her employer should carefully consider the expected cost and time frame to obtain patent protection, and if so, must act quickly to file a provisional or full patent application. If the inventor instead decides to pursue trade secret protection, it is critical that anyone exposed to confidential information about the invention be required to sign a confidentiality or non-disclosure agreement (NDA).

2. Trademarks and Domain Names

Before finalizing a product name or name-based logo, particularly one that would be imprinted, etched, or incorporated into the housing for a device, it is important to ensure that the product name and identifying information are clear for use. This typically involves trademark searches for freedom to use and protectability. Domain names can also be an important part of the process, but securing a relevant domain name will not provide value if the underlying trademark cannot be used.

Apple showed, with its release of the iPhone into markets where several other "iphone" trademarks and domains existed, that some types of trademark problems can be solved by creativity, ample resources, and negotiating leverage.

3. Arrangements with Developers, Contractors, and Licensors

Connected devices may have a multitude of design and engineering attributes, including industrial design, branding, circuit board design, power system design, RF engineering, ASIC design, firmware, other embedded software, databases, operating systems, and applications, to name a few. In many cases, the design and development of a new device may be outsourced or contracted to third parties, may be in-licensed from third parties, or may be procured from open-source or other public domain sources.

In any of these contracts, allocation of ownership of designs or innovations is critical. If no allocation of ownership is specified by contract, an outside contractor who creates work product (whether software, hardware designs, or patentable material) will likely be considered the owner of the intellectual property rights in that work product. In other words, if a company hires a design firm but does not have a written contract that specifies that the hiring company owns the work product, the outside design company will likely keep ownership of that work product. The hiring company's right to use the work product could be limited to an implied license.

If existing design elements, software, or other components are included in a new product, the use of these components may require a license from a third party. This may be true for either proprietary components or open-source components. Users of open-source components should review the relevant open-source license to ensure that the terms of those licenses do not have unanticipated effects on proprietary rights in the device. A common concern of design or manufacturing companies is that, under some open-source licenses, innovations or enhancements on the open-source framework must be made available to the public.

4. Labeling Requirements

Even small connected devices typically bear a variety of information, sometimes in obscure locations or buried in packaging, including serial numbers, country of origin markers, patent application numbers, SKUs, FCC and CE logos, and RoHS (EU Restriction of Hazardous Substances Directive) compliance designations, as well as licensed marks from Underwriters Laboratories (UL), Energy-Star, Bluetooth, the Wi-Fi Alliance, or others. These marks can be necessary for compliance with law, or may be designed to give assurance to a consumer as to the compatibility, quality, or safety of a device. Each of these marks may require the manufacturer to have undergone testing by an independent lab, certification, or licensing from a third party.

In addition, devices in some industries may require labeling, inserts, stickers, or warnings, because of regulations, local laws, or common liability limitation practice. Medical devices may require extensive labeling to comply with regulations. Wi-Fi access points sold in California must include a temporary warning sticker that must be removed by the consumer in order to allow the use of the device, with warnings about the dangers of unsecured networks. Some devices have safety labeling, or labeling targeted at avoiding specific kinds of accidents ("Not to be used for therapeutic purposes," "Not to be used for navigation of aircraft or operation of nuclear power facilities," and "Do not swallow" are examples).

5. Protection of Information and Data

Connected and location-aware devices, particularly those that cache, store, or transmit personal data, may be subject to claims under federal or state law, if the nature of data, use of data, and protection of data are not carefully and correctly described in a user-accepted privacy policy or other user document. Any loss of confidentiality, breach of security restrictions, or exposure of data can lead to millions of dollars of damage (on average, the Ponemon Institute estimates these damages in excess of $7 million per incident).

Worldwide, many jurisdictions, including the EU, have comprehensive privacy standards that can be violated by any transfer of user data, or collection of data from users, from the EU to a jurisdiction deemed to have more lax privacy standards (like the United States).

One type of data that has sparked interest recently is information regarding the location of wi-fi access points. Companies, including Apple and Google, have been interested in collecting data that correlate wi-fi access points with GPS information. With a comprehensive database of this information, these companies can provide location-based services relying on observations of wi-fi access points. This capability can greatly expand the number of devices to which location-based advertising could be served, among other things. While Apple had disclosed that the iPhone and iPad collect this information, and that the user had an option as to whether this information would be returned to Apple, users became outraged (and sparked a Congressional investigation, among other things) when they learned that this information is stored on the Apple devices in unencrypted form. The fear is that a thief or hacker could access this information and recover sensitive information about a user's location history.

Google has been sued worldwide because, in its collection of wi-fi access-point data, it also collected user information, including emails, passwords, and other information, transmitted by users across unencrypted wi-fi connections. Google has stated that this collection was accidental, and that it was quickly stopped.

In these two cases and others, device manufacturers have run into legal trouble even when the collection of data was disclosed, or was arguably publicly available, because of heightened sensitivity to the collection or transfer of information users consider private.

6. Warranty Terms

Basic product warranties provide that, if a product does not function materially in the way described in its documentation, the company that offered the product will accept return, and will replace the product or refund the purchase price. Many manufacturers ship products with an express warranty on a card that states that the manufacturer's only liability is for repair or replacement of the product, and that the manufacturer cannot be held liable for other damages that a device may cause. Offering an express warranty, and disclaiming other warranties, may also allow the manufacturer to reduce liability for certain warranties that are implied by state laws, such as a warranty of merchantability, fitness for a particular purpose, or noninfringement.

These terms can be particularly useful if a device could be implicated in damages that exceed the purchase price of the device, as when a device handles user data or is involved in vehicular applications. Faulty, inconsistent, or obsolete location-based data, of the kind that many connected devices utilize, has been implicated in situations ranging from automobile accidents to border disputes. As connected devices manage growing volumes of personal data and location-based data, manufacturers and distributors need to be fastidious about the way they construct legal terms that govern this data and work to limit liability.

7. Service Return and Recycling

Under a product warranty, a manufacturer will often set up a Returned Materials Authorization (RMA) process for accepting returns and processing repair or replacement of devices. For connected devices, the RMA process and any remote device service must be structured to minimize contact with user data, to disclose the collection, use, and protection measures applicable to user data, and to ensure that discarded or recycled devices containing user data are not made available to the public.

As concerns about the impact of technology waste on the environment have grown, so have regulations about how devices must be recycled or reused. Manufacturers and sellers may be required to register, collect surtaxes, establish recycling facilities or programs, and contribute to state funds for recycling. For many companies, the "throwaway" portion of the device lifecycle has become a significant budgetary item.

8. Export and Cross-Border Issues

The ready access to a global marketplace offered by the Internet can expose a device company to an enormous and inconsistent patchwork of export and import regulations, costs, and paperwork. Every jurisdiction where a customer lives may impose its own laws on those who ship products to customers in that jurisdiction. In addition, the device company's jurisdiction may regulate the outbound shipment of devices, particularly those with potential military applications or encryption capabilities.

Manufacturers who outsource development often ship or procure components, work-in-process, and finished goods across several borders. Customs clearance, taxes and duties, and regulatory compliance may be necessary at every step.

Bringing a product to market has never been simple. While the last decade's swift embrace of connectivity has allowed even small companies like Canary Wireless to reach a global marketplace, the challenges facing product companies have multiplied. Those who are drawn to the revenue streams that spring from connected devices must be prepared for adventure in navigating a landscape marked with legal, regulatory, and liability hazards.

Chapter 15

Speedy: In a New Era of Perishability

Charlie Feld is a highly respected figure in information technology circles. As an executive and a consultant, he has shaped the technology direction at Frito Lay (the snacks subsidiary of PepsiCo, Inc.) and Delta Airlines, among others. As he talks even about today's trends, you can see the influence the time at Frito Lay three decades ago had on him.

He describes conversations with a long retired Herman Lay, one of the founders. Lay described the business in very basic terms: "No matter how big or complicated the company gets, we still buy potatoes, cook them, put them in bags, sell them to our customers and collect the money."[1]

Feld also describes periodic "field trips" with route drivers who stocked Frito Lay snacks at various outlets. There were constant reminders that their product was perishable and no amount of sophisticated industrial engineering or computing could ignore those basic principles of nature.

Charlie Fine, an MIT professor, wrote a seminal book in 1999 where he introduced the concepts of "clockspeeds" and "temporary advantage."[2] He also suggested businesses learn from fast-paced industries as geneticists do from the short life span of a fruit fly.

A studio executive is discussing the challenges of demand forecasting in the business. How many BluRay versus traditional DVD copies of a new movie to ship to each city? How soon before we can plan for streaming video even for new releases? Then he confides that with the speed at which his industry is moving, he has considered recruiting candidates from the poultry farming industry.

"They are much more sensitive to short lifecycles."

Potatoes, fruit flies, poultry, and technology—welcome to the new world of perishability.

Time Really Hurries Faster These Days

The Flip camera went on sale in 2007 and quickly became a dominant camcorder brand. Cisco bought the company in 2009 and by then it had sold over 2 million units. Just as quickly, it shuttered the unit in 2011. Commented the *New York Times,* "Even in the life cycle of the tech world, this is fast."[3]

If that is fast, how about the fact that Apple gets about two-thirds of its revenue from iOS devices, a platform that didn't exist four years ago?[4] But analyze that a bit further and see how many generations of iPods and iPhones and iPads were launched (and predecessors deemphasized) in that time frame.

Or that the first billion applications downloaded on Google Android phones took 20 months, the second billion installs took another five months, and the third billion took only two months. Again, analyze how many Android devices have been introduced and deemphasized since the HTC Dream, the first commercially released Android smartphone in October 2008.

Rapid Product Iterations

Just look at how quickly navigation apps and devices have evolved. We were thrilled when basic GPS and maps arrived in our cars with GM OnStar. They evolved to allow for emergency services. The economics and portability of standalone GPS devices led us to Garmins and Tom Toms. Then we started to see GPS devices that could recognize voices and work even in tunnels.[5] Still others allowed Bluetooth calling and live traffic updates and rerouting. Then just as quickly, many of us moved away from dedicated GPS devices to Google Maps and other navigation apps on our smartphones.

Today Google has "self-driving" cars:

> Our automated cars use video cameras, radar sensors, and a laser range finder to "see" other traffic, as well as detailed maps (which we collect using manually driven vehicles) to navigate the road ahead. This is all made possible by Google's data centers, which can process the enormous amounts of information gathered by our cars when mapping their terrain.[6]

Soon, with the coming of electric cars and the need to charge them on a regular basis, we will need apps that not only monitor the battery level, but also allow us to make reservations at a charging station on the way.

All this has transpired within a decade, and we have rapidly gone from GM to Garmin to Google—and maybe back to GM? As we mentioned in Chapter 1, GM now sells OnStar FMV on a rearview mirror with many of its features to work even with non-GM cars. Toyota is working with Microsoft on a joint telematics offering for its upcoming electric and plug-in hybrid cars.[7]

Rapid Change in Competitive Landscapes

James Politeski, a Senior Vice-President at Samsung Electronics America, says in an interview, "As fridges and other home appliances get smarter and become more like computers that can connect wirelessly to smartphones, tablets, and other devices, Samsung is harnessing this transition to take business from appliance leader Whirlpool and other manufacturers,"[8] Whirlpool is used to competition from GE and Kenmore and globally from Electrolux and Haier—but a phone and electronics company like Samsung?

Jeff Brown, vice president of business intelligence for UBM TechInsights, wonders, "The question for traditional medical device companies is whether their designers, marketers, and IP staff have factored the smartphone platform into their thinking." He continues:

> Smartphones provide medical technology companies with unprecedented access to an enormous consumer market. To capture this opportunity, they must think carefully about how they develop new technologies and protect their intellectual property innovations. Otherwise, they face the same fate as makers of stand-alone GPS and MP3 players—a slow decline to obsolescence.[9]

Volatile Demand Forecasting

Demand forecasting has always been about working around peaks and valleys. So technology companies have learned to factor seasonal adjustments for back-to-school campaigns, "Black Friday," home entertainment around major sporting events, and so on. Now, try estimating demand for a brand-new product category for which there is no history to go by. Planning for sales of 25 million iPads in little over a year[10] or 2.5 million Kinects sold in the first 25 days[11] requires juggling between being stuck with too much inventory of perishable product and disappointing customers and opening their doors to competitive alternatives.

E-book readers have repeatedly had shortage issues. In the 2008 Christmas season, the *New York Times* wrote, "Now it is out of stock and unavailable until February. Analysts credit Oprah Winfrey, who praised the Kindle on her show in October, and blame Amazon for poor holiday planning."[12] In 2009, Barnes & Noble announced:

> Customers who pre-order the Nook now won't get the device until the week of January 4—after the holiday shopping season. A limited number of the devices will be available for sale in some of the "highest volume" Barnes & Noble stores.[13]

In summer of 2010, the *Wall Street Journal* wrote:

> But unlike with earlier (Amazon) Kindle delays, would-be customers now have more options if they want an e-reader now and don't want to pay $499 for the iPad. Without even an expected delivery date for the Kindle, they could turn elsewhere—to the $189 Nook from Barnes & Noble or to a plethora of devices from Sony and others.[14]

Nintendo is another company that has similarly had a tough time anticipating demand for its Wii and other game consoles.

Indeed, *CIO* magazine asked:

> Or is Nintendo purposefully withholding Wiis from U.S. customers—a shrewd marketing tactic to artificially create intense demand?
>
> "Conspiracy theorists are saying that since Nintendo has already met their end of March goals (with 6 million units shipped), and are building up supply, continuing the demand, and ensuring awesome second quarter sales," speculates a May 2008 Geeksugar blog. "Unfortunately, with American retailers now running at 2.5 percent availability, some are expecting the Wii shortage to continue through 2009."[15]

By the end of 2008, however, Nintendo of America president Reggie Fils-Aime was hedging, "The video game industry has weathered recessionary times fairly well . . . (but) if we get into unchartered territories with stocks coming severely down and unemployment spikes, then all bets are off."[16]

Amazon and Nintendo undershot. HP spectacularly overshot with its TouchPad and killed the product within a few weeks of announcing it. Both extremes are common in a world of volatile high-tech demand forecasting.

Moving from Physical to Digital Supply Chains

To get away from the vagaries and volatilities of the physical supply chain, Amazon has aggressively promoted software—its Kindle reader app that runs on PCs, Macs, iPhones, iPads, BlackBerries, and Android phones.

While that may cannibalize sales of its Kindle device or Fire tablet, it still benefits from not having to deal with the even hairier physical logistics of printed books. Amazon has cheerily announced its "ebook sales now outnumber physical paperback and hardback sales, the digital platform overtaking print in just four years." Since April 1, Amazon says it has sold 105 Kindle eBooks for every 100 print books on its U.S. site. That's not just limited to cases where both formats are available, either; it takes into account print titles where no Kindle equivalent is on offer."[17]

In ancient times, the Library of Alexandria was the greatest single archive of human knowledge. It has been estimated "that at one time the Library of Alexandria held over half a million documents from Assyria, Greece, Persia, Egypt, India, and many other nations. Over 100 scholars lived at the Museum full time to perform research, write, lecture, or translate and copy documents."[18]

The nonprofit Internet Archive is trying to recreate a modern-day Library of Alexandria with a collection that "includes texts, audio, moving images, and software as well as archived web pages in our collections."[19] That is quite an undertaking as "Many early movies were recycled to recover the silver in the film."

Devices like the Kindle are making it easier to access documents from the Archive. Amazon provides instructions to access "over 2.5 million free e-books to read, download, and enjoy," many of them out-of-copyright, pre-1923 books. With the Fire tablet it introduced in late 2011, it is positioned to provide similar access to videos, games, and other content.

Then there is the Google Library, which is meandering its way through courts around the world. Google has scanned roughly 12 million books from some of the country's finest libraries, in what it has said was an effort to provide easier access to the world's knowledge.[20]

Impact on Operations

With products evolving so quickly, can the back office—accounting, human resources, procurement—continue at its old pace? Can ERP global rollouts take years?

Look at Groupon, one of the fastest-growing companies in the world, which has added on average one new country to its operations every three weeks. Not surprisingly, it is in a hurry in every aspect of its business. In three months, 26 international markets went live on the ERP software, NetSuite, replacing hundreds of spreadsheets. Zach Nelson, NetSuite's CEO, explains, "The power and flexibility of the NetSuite cloud made it possible for Groupon to deploy global ERP in all those countries in three months instead of the three years a company of their size and trajectory would have faced with conventional, on-premise software." Nelson continues, "And the fact that the cloud was the only viable solution for the fastest-growing company of the millennium has not been lost on other enterprise customers. The world's biggest companies are going to the cloud, and quickly."

CEO and President Mark Symonds of Plex, another ERP vendor, talks about one of his customer projects:

> Inteva Products is a global tier-one automotive supplier that needed to replace its SAP ERP system and 30 other business systems in 14 global locations within a year. Plex Systems was able to accomplish the entire transition for Inteva's 1,200 users and 300-plus suppliers located in Germany, Mexico, Hungary and the United States within 12 months, a timeframe that would have been unattainable with a traditional on-premise ERP solution. In addition to quickly transitioning to cloud-based ERP solution Plex Online, Inteva Products was able to reduce its monthly IT costs by one-third.

Conclusion

A key trait of the new technology elite is speed in product innovation, in anticipating changes in competitive landscapes, in managing volatility in demand forecasting and supply chains, and even in the back office. In the next section, let's look at the Corning Gorilla Glass product, which defines the new clockspeed—it has been adopted by more than 400 electronic products in less than three years.

Case Study: Corning—The Gorilla® Glass Rocket Ride

> The Auto industry has four-year design cycles. Consumer electronics, in contrast, are moving to four-month cycles. And they need a much wider range of shapes and sizes. In 2007 we had one device using Corning Gorilla Glass. Now over 400 products have it.

Dr. Nagaraja Shashidhar (who goes as "Shashi"), business development manager at Corning, is describing a few of the demanding dimensions of one of the most successful product launches the 160-year-old specialty glass and ceramics company has ever had. In 2011, Gorilla in its third full year is on track to reach $800 million in revenue.

"Specialty Glass and Ceramics" may be a misnomer. Corning has repeatedly influenced technology markets. *The Generations of Corning*,[21] a book written in 2001 to celebrate the company's 150th birthday, says, "Very few companies have made inventions that have affected humankind profoundly. Corning has been involved with at least three: electric lighting, television, and fiber-optic communications. Glass is the hidden but essential material that makes all three work."

Now Corning can claim a fourth. Corning explains the functionality and the fashion appeal of Gorilla Glass as "Scrapes, bumps, and drops are a fact of life, but Gorilla Glass enables your device to resist damage from the abuses that come with everyday use. Gorilla Glass also has strong aesthetic appeal. It's thin, lightweight, and cool to the touch—enabling the sleekest designs."[22]

American Tourister made a name for its hard-shelled luggage with commercials that showed a gorilla throwing around its products. Corning is bringing out the gentle, protective side of the gorilla.

Continues Shashi: "We keep getting customer stories like the one whose device got run over by a truck. The device was a write-off but our glass survived intact!"

That magic calls for a chemistry lesson. Ion exchange is a chemical strengthening process where large ions are "stuffed" into the glass surface, creating a state of compression. Gorilla Glass is specially designed to maximize this behavior. The glass is placed in a hot bath of molten salt at a temperature of approximately 400°C. Smaller sodium ions leave the glass, and larger potassium ions from the salt bath replace them. These larger ions take up more room and are pressed together when the glass cools, producing a layer of compressive stress on the surface of the glass. Gorilla Glass's special composition enables the potassium ions to diffuse far into the surface, creating high compressive stress deep into the glass. This layer of compression creates a surface that is more resistant to damage from everyday use.[23]

It also calls for a legal lesson. "We have device-specific NDAs (nondisclosure agreements) with many of our customers," says Shashi, which means that Corning employees wouldn't be able to confirm all brands that use Gorilla Glass.

Market analysts speculated after the tsunami whether many popular handheld and tablet devices were made at Corning's Shizuoka, Japan, plant or came from a Japanese competitor, Asahi Glass. In Walter Isaacson's biography of Steve Jobs, he describes a meeting between Jobs and Corning CEO Wendell Weeks about Gorilla Glass for the iPhone. Corning, presumably due to a nondisclosure with Apple, would not confirm whether iPhone or other Apple products use that glass.

On the other hand, just as Intel had its successful "Intel Inside" branding, Corning has moved to a "Gorilla Outside"-type branding campaign, and products like the Samsung Galaxy tablet and the Sony Bravia HDTVs proudly market their Gorilla Glass feature in their ads.

Corning—An Impressive Institutional Memory

Corning's website says "it is strengthened by history." Many of us grew up with Corning's consumer products including brands such as Pyrex, Corningware, Corelle, and Revere. *The Generations of Corning* documents well much of the company's rich innovation history.

Corning spun off its consumer goods division in 1998. Today, it mostly focuses on industrial markets such as:

- Ceramic substrates and filters for mobile emission control systems.
- Optical fiber, cable, and hardware and equipment for telecommunications.

- Optical biosensors for drug discovery.
- Advanced optics and specialty glass solutions for a number of industries.

Corning has not, however, forgotten its consumer roots. Today Corning makes glass substrates for LCD flat panel televisions, computer monitors, laptops, and other consumer electronics.

Corning has repeatedly shown an impressive memory and ability to go back into its decades-old vaults and pull out technology to create products as markets mature. An example: Dr. J. Franklin Hyde, a Corning chemist, invented a process in the 1930s for making an almost-pure glass used today in fiber-optic technology. A market for that product only took off in the 1990s as telecoms internationally started moving away from copper in their networks.

Gorilla Glass has its own story of such institutional memory. In the 1960s, Corning launched "Project Muscle." According to company lore, then-President Bill Decker told the Research Director, Dr. Bill Armistead, "Glass breaks. . . . Why don't you fix that?"

Project Muscle led to a new ultra strong glass material called Chemcor, designed to stand 100,000 pounds of pressure per square inch. There are films of it being bombarded with frozen chickens and it would not chip, let alone crack and break.

The Generations of Corning documents that there was plenty of excitement about Chemcor prospects. It showed up as the windshield in Ford Mustangs, in spacecrafts, and even as unbreakable tableware. While the glass did well in specific markets like aircraft cockpit windows (to survive bird strikes, among other threats) and windows for jails, it was *too* strong for many applications and never truly took off as expected in the 1960s. Variations of this glass family continue to be manufactured in a small scale for ophthalmic applications.

In 2006, a Corning New Business Development team saw that cellphone covers offered a new opportunity for toughened glass.

This is not the old Chemcor. As Shashi points out. "We drew on the company's prior expertise with strengthened glass. However, Gorilla Glass is a different product and glass composition than Chemcor."

Gorilla Glass draws on Corning's proprietary "fusion-draw" manufacturing process. Molten glass is fed into a trough called an "isopipe," overfilling until the glass flows evenly over both sides. It then rejoins, or fuses, at the bottom, where it is drawn down to form a continuous sheet of flat glass.

In order to get to that combination of high-volume scale, unique attributes required for cover glass applications, and exceptional thinness and sheet quality, the R&D team had to develop a new glass formula that would allow them to do the fusion-draw production in the Harrodsburg, Kentucky, plant instead of the slot-draw production the project had started with in the Danville, Virginia, plant. The Harrodsburg facility has excelled in its role as Corning's glass-melting technology center, where engineers work closely with scientists at the company's Sullivan Park (Corning, New York) R&D Center to develop advanced fusion-formed glass for display, photovoltaic, and other emerging applications. From February through May of 2007, the Corning R&D and manufacturing teams raced through trials and research to develop the new composition and manufacturing process that, amazingly, worked on the first try.

The Missionary Selling and the Ramp-Up

Even though Corning knew it had a winner, it took plenty of customer coaching to think about glass, instead of plastic, in their product design in the 2007 and 2008 time frame. Shashi talks about several customer visits where Corning taught the basics of strength of glass and how to design components with glass, affectionately called Glass 101. Part of the education also involved visits to the famous Corning Museum of Glass, where industrial designers could get inspiration to design products with glass. The Corning Museum of Glass is an entity legally separate from the company and does not specially feature Gorilla Glass or related products.

Glass is supposed to be brittle (as President Decker had expressed much earlier), so one of the most impressive sales techniques to combat that image is the ball-drop test—an industry standard test for glass durability. A 1.18-pound iron ball is dropped from 1.9 meters. The Gorilla Glass gives and then returns to its original position—still in one piece. At trade shows attendees were invited to scratch, pierce, and otherwise torture a piece of Gorilla Glass. It survived just fine.

The social buzz started growing as YouTube videos of the ball drop and the trade show tortures started circulating. Corning also put out its own "A Day Made of Glass" video, which shows futuristic architectural, automotive, 3D TV, and large-panel displays such as highway signs, all made from glass. That is one of the most-watched corporate videos ever produced.

Consumers started asking manufacturers whether their displays were made from Gorilla Glass. Manufacturers also saw the payback. Dr. Donnell Walton, worldwide applications engineering manager for Gorilla Glass at Corning, said,

> Manufacturers want their customers to be satisfied with their devices—they don't want people returning a broken or scratched product. The damage resistance of Gorilla Glass along with the glass expertise of Corning is the reason why we're designed into hundreds of products in the market today.

In 2009, 14 devices used Gorilla Glass. By 2010, over 350 used the product!

That, of course, has had significant manufacturing implications. Corning has significantly expanded its plants in Harrodsburg, Kentucky, and Taichung, Taiwan. It has retrofitted portions of a liquid-crystal display (LCD) plant in Shizuoka, Japan, to produce Gorilla Glass. Loading, securing, tarping, and transporting glass poses logistical challenges. The Asian plant decisions were made easier by that logistical reality and the need to be closer to many of the device manufacturers in Asia.

At the Annual Shareholder Meeting in April 2011, Jim Steiner, senior vice president, Corning Specialty Materials, told the audience that Corning had improved the glass composition three times since launch and that it had applied advanced finishing technology to improve product performance.[24]

Such continuous improvement is key, as Corning chases newer markets with the additional manufacturing capacity.

Gorilla Glass in TVs and Applications Beyond

The biggest short-term opportunity is in next-generation, frameless, no-bezel TVs with edge-to-edge glass surfaces. There would be other advantages beyond the sleek look. A 55″ LCD TV using Gorilla Glass "is 8.5 lbs. lighter"[25] and "its exceptional durability provides damage resistance against bumps."[26]

"The sleek and sophisticated look of edge-to-edge TV designs is made possible by using a thin, durable, crystal-clear glass cover. Gorilla Glass provides the perfect solution," says David Loeber, business director, TV Cover Glass. "It's an outstanding cover glass—lightweight and sophisticated, while protecting against everyday wear."

Next there are automobile applications. Hyundai incorporated Gorilla Glass in its next-generation electric concept car, the Blue2.[27] There are opportunities in the "kitchen of the future" as appliances incorporate larger displays. CollinsWoerman, a Seattle architect firm, envisions a role for Gorilla in its vision of a pre-fab high-rise building.[28]

The glass itself keeps evolving. At the January 2012 Consumer Electronics Show, Corning announced version 2, which is 20 percent thinner and allows for even more sleek devices and brighter images. Corning is accustomed to boom-bust cycles when it comes to technology markets. The 2001–2003 technology meltdown slowed down fiber optic demand. More recently, growth in demand for LCD TVs, another big Corning market, has slowed. Even if the newer TV, auto, and other applications do not materialize, what Corning has already done with Gorilla Glass in a short time frame is impressive.

Prof. Charlie Fine at MIT says we can learn from fruit flies as we evaluate technology clockspeeds. Corning shows we can also learn from Gorillas.

Chapter 16

Social: Amid Chatty Humans and Things

Guy Kawasaki is a well-known brand in tech circles. He was part of the early Apple gang involved in marketing the Macintosh in 1984. He has since written 10 books and been a speaker at a number of technology events. He used social media extensively in launching his last book[1] and is a firm believer that it can be leveraged for any product launch, not just a book launch.

Kawasaki used a Facebook Fan Page, a book-specific website, Twitter tags, optimized search words, and email campaigns; shared review copies with influential and not-so-influential bloggers; and held photo contests and quizzes. The total budget for all these activities, he says, was under $20,000.[2]

At the other extreme, a 30-second ad on the U.S. football Super Bowl event costs plenty. It was estimated at $3 million for the 2011 game. The cost reflects the fact that in most years it is the most watched TV show across the world. The 2011 game drew an audience of over

110 million, even in countries where football means something quite different and the game starts at ungodly hours. The production of the commercials costs even more. You are talking real money here compared to Kawasaki's budget above. Yet in 2011, the underlying theme for most of the Bowl ads was integration with social media.

The biggest change from previous years: The commercials did not point to advertisers' static websites, but instead to their Facebook pages, Twitter accounts, or iPad apps.

Big budget or small, it is clear that social savvy is a required skill in the new world, especially when it comes to product launches.

Super Bowl and Social Media

Social Strategy1, a firm that tracks media intelligence, compiled data on social channels around the 2011 Super Bowl. It found "Twitter was the dominant game-day source of social media activity. Viewers who were mobile-enabled seem to prefer Twitter for real-time conversations. YouTube and other sources had much more impact the week following the game, although Twitter was a strong player in both periods."[3]

Advertisers on the Bowl blended traditional and new media in creative ways:

- Mercedes-Benz ran a "Tweet Race." Four teams raced to Dallas (where the Super Bowl was being held) in Mercedes cars, fueled by how many times fans tweeted using the hashtag of their celebrity-led teams (Serena Williams, Rev Run, Nick Swisher, and Pete Wentz) from Chicago, New York, Los Angeles, and Tampa in specially equipped Mercedes vehicles.
- Volkswagen's ad, which featured a young Darth Vader discovering the "Force" around its Passat, was one of the most viewed on YouTube after the game and was the "most loved" in a survey of tweets.[4]
- Perversely, Groupon, which is obviously web/social savvy, as we have discussed in other chapters, created plenty of buzz, albeit negative, with its Tibet-focused commercial. A sample Tweet was "That

@groupon commercial? 50% off having a clue how to spend millions on a Super Bowl spot."

Beyond the Super Bowl, the integration between TV and radio commercials and social media is accelerating. Paramount Pictures, while launching *Transformers 3* in the summer of 2011, allowed fans to "Shazam" TV and radio ads for a free download of an exclusive live version of the Linkin Park single "When They Come for Me." Fans could also watch a featurette, buy show tickets, or purchase the soundtrack.

Shazam, which calls itself the world's leading mobile discovery platform with its reach of 125 million users, has been used by Procter & Gamble, Starbucks, Honda, and other major brands to integrate traditional advertising with mobile sites.

Swag and Social Impact

Google's annual developer conference I/O usually has plenty of announcements. In 2011 it was about Google's streaming music, Chrome, and Android momentum. The real eye-popping watermark the conference keeps raising year after year is the swag ("stuff we all get") that attendees qualify for, and then talk, tweet, and blog about:

> Google I/O 2008 had one Keynote speech, and Android phones were a presentation, not a giveaway. For Google I/O 2009, every attendee received a "Google Ion" phone, which was a special edition HTC Magic running Android 1.5. [In 2010], every attendee received one of two Android phones, a Nexus One, or a Motorola DROID, before the conference even began. Then at the second day keynote, they were told they'd get a not-yet-released-to-the-public HTC Evo 4G as well.[5]

The 2011 event brought attendees a then-unreleased Samsung Galaxy Tab 10.1 Android tablet, a Verizon 4G/LTE mobile hotspot, a developer version of the LG Optimus 3D (a 3D glasses-free smartphone), and a Sony Ericsson Xperia Play smartphone.

For 5,000+ attendees that may seem like a big marketing expense, but given the influence and social reach of these attendees in the technology early-adopter market, it is often a wise investment.

Cirque du Soleil offered free tickets to its show to bloggers during Blog World in Las Vegas in exchange for an "honest review on your blog or podcast." Said a blogger (on a blog, of course): "I think that this was a brilliant idea. What better advertising could you get than reviews from bloggers, some of whom have a network of hundreds or even thousands of subscribers?"[6]

Dell's "Free-Range Marketing"

"I bumped into Michael Dell at All Things D after his interview, and he was nice enough to show me this laptop that he was carrying that he said no one's seen before. It's a small form factor notebook . . .," blogged Brian Lam about the mini-notebook he had just seen. [7] It turned out to be a yet-unannounced Inspiron 910 (also known as the Mini 9).

Sometimes, the social buzz is unintended (though Mr. Dell is certainly savvy enough to have leaked the story by carrying a hard-to-miss red-colored device) and it helps for a company to move quickly.

Dell's official launch for the notebook was more than four months after the original blog post on Gizmodo:

> They needed to be mindful of their upcoming launch activities, so they turned to what they termed "free-range marketing": allowing the community to drive the excitement and the story about the new product. Blogs, forums, and social bookmarking sites like Digg were abuzz with talk about the new Dell notebook.
>
> Dell actively monitored the ongoing dialogue, absorbing the information and feedback from the community using tools from Radian6.[8]

Radian6 has since been acquired by Salesforce.com, and is a key component of the Social Marketing Cloud the company announced in December 2011.

IT and Brand Impact

IBM, which certainly knows about IT audiences, used its Lotusphere conference in early 2011 to announce that we are seeing the "fifth shift in business technology"—from mainframe to departmental computing to the PC to the Internet and now to social business.

Bruce J. Rogow, introduced in Chapter 5, puts it even more crisply:

> 20–70 percent of a firm's brand is now delivered or supported by IT.
>
> Historically, a firm's brand and brand equity was built through advertising, promotion, and the customer's confident use of the product or service. Today, the brand equity ebbs and flows with the quality of the call center, the ease of use of a website, ability to properly deliver a product on time, and the ability to have the right version of the product where it should be. The coolness factor includes the apps that are available to support the product or the buzz created in the social media. That buzz can be positive or negative. Who hasn't seen the YouTube video "United [Airlines] Breaks Guitars"?
>
> My new Audi A4 broke down. Cars break down all the time, of course. But it took them nearly a month to locate the proper part, figure out what was wrong, and get the expertise to fix the problem. I posted over 25 missives on Facebook, Twitter, and the like. In today's properly IT-supported business it should not take a month. The brand suffers.
>
> In the B2B space, companies expect that their suppliers can seamlessly integrate into their automated supply chains. They expect the suppliers' inventory, pricing, availability, ordering, refresh, return, product information, and invoicing to be world class. The brand is no longer just about having your decal on Jeff Gordon's NASCAR or having your blimp fly over a football game. Most customers now experience your firm's real brand through interaction or lack thereof with your IT-supported vehicles.

No matter what Rogow says, many IT leaders are still cynical about social networks. Bill Brenner is an editor of *Chief Security Magazine,* so he is conflicted between attracting new readers via social channels and his job of informing his readers of technology security risks. He shared his ambivalence as he joined Google+, the social network described further on. "But the business value is worth the risk for me. Facebook, LinkedIn, and Twitter have been a big source of new CSO readers, and I expect Google+ to help us further. As long as I don't decide to actually trust these sites, I figure I'll be okay. I hope."[9]

Many CIOs are not ambivalent. They see little upside from social networks. Indeed, some surveys suggest almost half of them see such networks as employee time wasters. The irony is that their own HR colleagues, whose job is to worry about employee productivity, have embraced social networks like LinkedIn.

Marc Andreessen, co-founder of Netscape, one of the first browser companies, and now a general partner of the venture capital firm Andreessen-Horowitz, puts it succinctly:

> LinkedIn is today's fastest-growing recruiting company. For the first time ever, on LinkedIn, employees can maintain their own resumes for recruiters to search in real time—giving LinkedIn the opportunity to eat the lucrative $400 billion recruiting industry.[10]

"Gamification"

Gabe Zichermann is an author, event organizer, and blogger around an exciting new trend called "gamification"—the use of games and rewards to engage users. He provides his first example:

> With Nike+, Nike's goal was obviously to generate brand loyalty and ultimately sell more sporting equipment, but it clearly thought very carefully about what kinds of people would use the application and prioritized those players' needs first. It didn't simply start assigning points and badges for buying Nike products; instead, it sought to make running more fun and thereby attract a large community of runners to whom it could then market Nike products.
>
> A player can jump right in and begin using the app as little more than a pedometer with a stopwatch. As a newbie, she can begin by playing against herself, competing against her best time or best distance, using a leaderboard of her runs to motivate her to keep improving. But as she continues to explore and use the application, new games are presented.
>
> Nike+ adds a social layer to the basic run-tracking game, creating a much richer experience for its players. Runners are encouraged to connect to Facebook and post their run information to their feeds. When a player begins her run, the app posts a notice to her Facebook feed and asks her friends to cheer her on. Each time a friend "likes"

the post, the app plays a burst of roaring crowd over her music, to let her know that a friend has just supported her efforts. This opens up a fun social loop that reinforces the player's commitment to her fitness program, whether she is training for a marathon or going for a casual jog. At the end of her run, the player can see the supportive comments her friends posted on Facebook. But the app also includes surprise encouragement and positive feedback from celebrities such as Lance Armstrong and Tracy Morgan, adding a variable reinforcement touch. Beautiful "heat" maps show her where she was running fastest and slowest, making it more fun and engaging to review a run.

The core application is a handy tool for runners to track the time and distance of their runs. But the skillfully employed game mechanics take this basic pedometer and turn it into something far more social, engaging, and fun, subtly drawing the player into the game and making her want to come back again and again."

Zichermann continues with another example:

Jigsaw (acquired by Salesforce.com) is a business-to-business marketplace for contact information that has successfully employed a leaderboard in order to encourage the uploading of consumer data. Essentially a wallet for virtual "business cards," Jigsaw allows people to both upload and download full contact information for anyone. Although somewhat controversial—you can upload anyone's contact information with or without their consent—the idea is simple enough, and quite compelling. However, it would be impossible to create an up-to-date database of 16+ million professionals' contact information in a cost-effective manner—so Jigsaw relies on users to upload the business cards they gather at trade shows, conferences, and through their social networks.

But one of the site's most compelling strategies is its leaderboard—a place where users can compare their performance within the network against each other. In fact, the company offers leaderboards for each of the core activities it's seeking to promote: New Contacts Added, Contacts Updated, Referrers, and Wikis (notes). These leaderboards (and the challenges, levels, badges, and points that they summarize) have enabled the company to move from cash compensation to virtual compensation, eliminating huge costs, gaining substantial advantages in their competitive space, and encouraging users to take actions that directly benefit the company without any obvious reward.

For some users, their ranking on the leaderboard is their reward.

For other players, and possibly most importantly for marketers, the greatest power of leaderboards is that they are also a decisive indicator that some kind of game is being played. They reveal the existence of a ranking mechanism, signal the existence of a point system, and suggest the presence of rules for ways to garner those points. They are an essential part of the puzzle in game based marketing.

Other creative uses of gamification include:

- When it launched the HTC EVO 4G phone, Sprint offered up "virtual badges" when users perform EVO 'firsts' including tweet, upload an EVO unboxing video, check in on Foursquare, or take the 'first photo of an eight-person 4G hotspot' (a unique feature of the EVO) using their device. Those badges—which prominently feature "4G" images—can then be shared on Twitter and Facebook. Users who are 'first' get featured on the EVO homepage too."[11]
- Warner Bros. used Foursquare to promote the release of the movie *Valentine's Day* starring the (social media savvy) star Ashton Kutcher. The promotion involved "Romantic Tips" around the cities (New York, Chicago, San Francisco, Los Angeles, and Boston) that the movie featured. If users checked into any two of these locations, they unlocked a special Valentine's Day badge.

The Digital Crowd Queuing around the Block

We have seen the lines that form outside Apple stores around the world when a new product becomes available. Something similar happened virtually when Google+ was launched in July 2011.

Google+ is Google's response to Facebook with the differentiation of "The easiest way to share some things with college buddies, others with your parents, and almost nothing with your boss." And it was invitation-only for its "Limited Field Trial."

The scarcity of invites led to a dramatically increased social buzz and even auctions of the invitations on eBay. There were also scams.

Cyber security firm Sophos exposed a Google+ invite scam that sent out invites that look like the real thing, except the link took the tricked users to a pharmacy web site that sells Viagra.[12]

John Quelch, marketing professor at Harvard Business School, wrote:

> Marketers are trained to match supply to demand. Everything that consumers need should be available at the right time in the right place at the right price. Coca-Cola's mantra always has been to be within an arm's reach of desire. To be out of stock is to lose a sale or, worse, to lose a sale to a competitor.
>
> But marketers also understand that, by using the illusion of scarcity, they can accelerate demand. This false scarcity encourages us to buy sooner and perhaps to buy more than normal.[13]

Google+ certainly created that virtual crowd waiting for the launch with those in the front of the line wearing "jealous?" T-shirts. It claimed 20 million members in the first month. Impressively, its stock reflected that excitement, and the market cap went up 30 percent, or $45 billion, in the same time period.

Yes, that's one heck of a crowd around the block.

"Things" Can Be Social, Too

Think social buzz only works with humans?

Japanese automaker Toyota is planning a private social network called "Toyota Friend," powered by a tool called Chatter from Salesforce.com. It will connect Toyota customers with their cars, their dealership, and with Toyota:

> Toyota Friend will provide a variety of product and service information as well as essential maintenance tips, creating a rich car ownership experience. For example, if an EV (electric vehicle) or PHV (plug-in hybrid) [both due in 2012] is running low on battery power, Toyota Friend would notify the driver to recharge in the form of a "tweet"-like alert. In addition, while Toyota Friend will be a private social network, customers can choose to extend their communication to family, friends, and others through public social networks such as

Twitter and Facebook. The service will also be accessible through smart phones, tablet PCs, and other advanced mobile devices.[14]

Japan, of course, also leads the world when it comes to social robotics. After its 2011 tsunami, many were soothed by Paro, the robot seal. The robot is covered in tactile sensors and responds to petting by squealing. The country is also experimenting with nurse robots and, in a decade or so, it is expected that instead of its traditional focus on industrial robots, two-thirds will be "service robots" used in care of its aging population.

With billions of sensors talking to each other and to us there is a parallel social network, broadly called the Internet of Things, we have yet to fully understand or leverage. A relatively simplistic application is that of shared washing machines and dryers that text and tweet students in a dorm, when their clothes are done.

Conclusion

The technology elite know how we live in a world of plenty of chatter, human and non-human. Learning to interpret that chatter and magnify it via social savvy is no longer a "nice to have." Not only your employees, but also your products need to be socially savvy. Let's next look at a socially savvy product, the Lexmark Genesis.

Case Study: Lexmark Genesis—A Printer for Our Social Times

"Taller than it is wide."

"High-gloss black piano finish."

The striking physical looks of the Lexmark Genesis are what every initial product review focused on when it was introduced in October 2010. Seventeen inches tall, it towers over every other device on the desktop or on the office floor. If the height does not get your attention, the glossy, sloping black front surely does.

Next you notice the 4.3-inch touchscreen LCD display, which is about the size of the smartphones most of us are used to these days. That's your gateway to the setup (who looks at user manuals these days?), ink levels, and the applications. Large, easy-to-read graphic icons drive the copy, scan, and fax functions. It also comes with "SmartSolutions," some pre-packaged and others downloadable from the Lexmark online SmartSolutions library.

Then you notice the blazing speed of the scanner.

Wrote *Engadget*:

> 750 milliseconds after you close its front-facing scan bay, the CMOS sensor generates a preview on the 4.3-inch color touchscreen, and 2.2 seconds after that, it's got a full 300 dpi image saved on your USB-connected computer or winging its way across 802.11n wi-fi.

It makes easy the choice to store the scanned documents as JPEG, PDF, editable Word document, or auto attach in those formats to e-mail. The Genesis packages a 10-megapixel digital camera instead of using charge-coupled device (CCD) arrays found in traditional scanners. The speed has led Lexmark to call the printer a "Now-in-One," a play on the traditional All-in-One designation for multifunction printers.

Finally, you notice the ease with which you can set up your home or office "cloud" as you can use its wi-fi connection for laptops, iPads, and iPods. Or you can use a port on its side to print directly from a USB device or memory card. That negates the need to upload pictures onto your PC and then print. While Lexmark SmartSolutions are designed for business, they can also be social, allowing users to scroll through and search their Twitter feeds as well as view their Facebook walls.

And why not? When everything from shoes to pens to grills is becoming smarter, why not a dramatic improvement in the boring old printer?

From Lexmark? The Kentucky-based company often does not get the exposure it would if it had been based in Silicon Valley, but it has a 20-year track record of innovation.

The Company

Lexmark was formed in 1991 when IBM divested its printer and related supplies business. In 2010, Lexmark sold products in more than 170 countries and reported more than $4 billion in revenue. It diversified into global manufacturing and software expertise over a decade ago. It has talent pools and manufacturing capabilities in Cebu, Philippines; Shenzhen, China; Kolkotta, India; Juarez, Mexico; and Budapest, Hungary.

It can claim a series of milestones such as Operation ReSource, a free laser-cartridge recycling program for customers way back in 1991. It can also boast the Medley, which was in 1995 the industry's first printer-fax-copier-scanner combination product with color printing capability, and in 1997 it introduced the industry's first inkjet device capable of printing 1200 x 1200 dpi. In 2006, Lexmark introduced Fleet Manager, allowing channel partners to deliver Lexmark's managed print services tools and technologies to their customers. By 2007, it offered customers the widest range of wireless printers. In 2010, Lexmark acquired Perceptive Software to expand its document workflow solutions portfolio.

In addition to products under its brands, it also white-labels printers sold by Dell and IBM.

Printers as Maligned Devices

Even with all the innovations in printers that Lexmark and its competitors like HP and Canon have delivered over the years, the printer is considered a frustrating device by most consumers and businesses. Blame that on frequent paper jams and other service issues, on expensive toners and other supplies, on long wait times to warm up/print/scan, for polluting the environment, and for archiving/shredding headaches of printed paper. In fact, retailers often use printers as loss leaders to sell other equipment.

Indeed, it is striking that *Consumer Reports* in its "Printer shopping tips" mostly emphasizes considerations on how to minimize usage:

- Avoid blank pages.
- Print fewer pages.
- Conserve ink or toner.
- Power it down.
- Seek efficiency.
- Recycle cartridges.

So, the Genesis design team had its work cut out for it. How do you get consumers excited about a significant form/factor leap in a device with a boring and frustrating image?

The Genesis Design Process

Robert Eskridge, in the Lexmark Product Development Engineering group, describes the design process for the Genesis:

> The key element in the entire process was innovation in how customers use our products and what customer benefits we provided at each decision level. That may sound obvious but required a disciplined multistep process.
>
> We started with early "white-paper" concepts. As with any new product, we looked at marketing input based on the review of customer feedback on our earlier products. Out of that, we knew the Genesis needed to be a smaller footprint and that guided us to the vertical look. We also looked at the competitive landscape and the current state of "the art of the possible" in terms of technologies.
>
> Next we got input from a handful of "Outcome-Driven Innovation" (ODI) sessions. ODI is based on the premise that when it comes to innovation, the job, not the product, must be the unit of analysis.

Based on conceptual mockups, we got customer feedback on how they used various copying, faxing, scanning functionality. This really helped with our Flash Scan decision as we saw customer delight with "document to digital" in three seconds compared to the much longer time they tolerate with scanning today.

We then did in-house testing and tweaking that covered the whole customer experience—unboxing, setup, ease of cartridge installation, networking, and using the device. One feedback point given the vertical platen was to engineer a card clip to hold in place smaller receipts, business cards, and photos.

We then performed online testing to validate various concepts (images, feature/function, pricing). I would estimate we had over 2,000 participants over the product design cycle. Finally, we moved to betasphere testing with product placement in customer environments.

A deep dive color study was done to determine the two models that we have today. One has a high-gloss black all around and the other has silver panel accents. No major technical changes were uncovered during the last few test cycles; however, tweaks on usability and messaging were implemented based on customer feedback.

The SmartSolutions

Kathy Edwards, in Lexmark communications, explains the concept of SmartSolutions, which are such an important part of the product:

They are focused on making our customers become more efficient. Most solutions were designed to help small and medium-sized business be more productive, thus saving time and money. Complicated tasks are made easy with one-touch workflow solutions.

Take a personal productivity example. I am always copying invoices and then filing them. With SmartSolutions, I can customize an App, name it Invoices, then scan every invoice into my Invoice File Folder on my PC or Mac. I can shred and recycle the hard copy, allowing my business to have everything in an electronic format for archiving, and I've become a little more efficient instead of spending time on copying and filing. All with one touch of a button. The solution was customized to fit my specific needs during my initial setup.

> Or take expense tracking. Businesses have charges and travel receipts that need to be maintained for tax purposes. With SmartSolutions, they can customize an App, name it Tax Receipts, and scan those copies to a file folder, or even better scan directly to their tax accountant via email. Again, the solution (scan to file or scan to email) was customized during their one-time setup.

Lexmark has a SmartSolutions store (like the Apple App Store or Android market) with a number of free and purchasable applications from Stamps.com, Evernote, Box.net, LegalZoom, and TripIt, among others.

The Social Product Launch

Jerry Grasso, Vice President of Corporate Communications, says the company leveraged social media extensively for the product launch:

> We launched the Genesis product at BlogWorld and New Media Expo in Las Vegas in October 2010. First time we have done that. We usually launch products at CES or other technology shows but over the last year we have seen a tipping point for the payback from leveraging social media. We launched simultaneously with press events in New York and San Francisco and even in those cities we invited a mix of bloggers in addition to more traditional media like the *New York Times*.
>
> Our analysis shows over 50 blogs or articles were written about Genesis within a few days of the launch with a readership of over 32 million around the world. You can measure quite easily the Tweets and other readership metrics from the write-ups in blogs like *Engadget* and *Gizmodo*. We found it was especially influential in the early adopter customer base.

The Genesis is definitely a printer for our increasingly social world. Its social features, and its social launch, give it a solid chance to reshape the worldview of printers.

Chapter 17

Sustainable: Mining the Green Gold

Many new cars in the United States, especially in the state of California, sport a sticker with their Global Warming Score. It's a score that ranks each vehicle's CO_2-equivalent value on a scale of 1–10 (10 being the cleanest) relative to all other vehicles.[1] The score is based on a calculation of various greenhouse gases, including carbon dioxide, methane, and hydrofluorocarbons from the air conditioning system, the car is likely to emit. The thinking is that over time consumers will increasingly factor this score in their buying decisions.

Sites like GoodGuide[2] score a variety of cell phones on a number of factors such as:

- Energy management, as measured by a phone's standby power consumption.
- Materials management, specifically the use of eco-materials in a phone or its packaging.

- Toxic waste, specifically inclusion of polyvinyl chloride or brominated flame retardants in a phone.
- Product management, as measured by third-party certification standards.
- Environmental disclosure, as measured by the availability of an environmental fact sheet for a product.

A company called TerraPass, which facilitates trading in carbon offsets, has long allowed consumers to calculate on its website the carbon footprint of their car model. It goes further and allows you to calculate the carbon equivalent of air trips you take and your home's carbon footprint.

These scores and ratings highlight that consumers are sensitive to the fact that our modern technologies like automobiles, aviation, and air conditioning all contribute to carbon proliferation. We can argue about whether that leads to global warming, but if we are building next-generation, smart products, should we not also be looking at efficiency when it comes to emissions? Since a dip in 2009 caused by the global financial crisis, the International Energy Agency reports that emissions are estimated to have climbed to a record 30.6 Gigatonnes (Gt), a 5 percent jump from the previous record year in 2008.[3]

Tony Prophet of HP, whom we met in Chapter 2, says it is a source of personal pride for him that the HP supply chain is sustainable. The comment is noteworthy given the dimensions of the HP supply chain are already extremely complex. It goes way beyond tackling carbon and other emissions from its products to those in its supplier operations and product packaging. HP's numbers are impressive—1.4 billion KWH of electricity customers saved from 2008 through 2010 using high-volume HP desktop and notebook PC families; recycled over a billion ink cartridges using HP's "closed-loop" process, which uses plastic from returned cartridges to make new ones; over 2 billion pounds of other electronic parts recycled since 1987.[4]

Prophet also talks about fair labor practices around the world and sourcing of "conflict minerals." He says that at a time when demanding analysts like "The Greenmonk" expand the definition of "sustainable" for the high-tech industry.

"The Greenmonk"

Tom Raftery has been pushing for sustainability for a long time. An Irishman who lives in Spain, Raftery bleeds green. After a stint at the Cork Internet Exchange, a "hyperefficient data center," he is now an analyst at Redmonk and looks for environmental improvements across many industries, but particularly in technology. Typical of his comments is this one:

> Facebook has a very efficient data center in Prineville, Oregon [as we discussed in Chapter 8]. Its PUE is 1.07, which is near the theoretical maximum (of 1.0), but it is powered by Pacific Corp, 63 percent of whose electricity is generated by burning coal—very definitely not Green. Same with Microsoft's Dublin, Ireland, data center—again a very respectable PUE of 1.2, but run on the Irish electricity grid, 87.5 percent of which comes from fossil fuels—again, not Green.

He invokes William Stanley Jevons, who published a book, *The Coal Question,* in 1865. The Jevons paradox says that as we become more efficient in use of materials, we just increase consumption. Raftery agrees: "Amazon Web Services allows anyone with $10 to rent a 10-machine cluster with 1TB of distributed storage for eight hours. Economically efficient, but in the big scheme of things, not very Green."

Surely, there are at least a few technologies he considers Green?

> I guess Smart Meters would be the first one. While not hugely sustainable in and of themselves, properly rolled out with Smart Grid technologies and demand management, they have the potential to greatly increase the penetration of renewables on the electrical grid and reduce global emissions.
>
> I love LED lights. I have a bunch of them installed around the house and they are fantastic. Great light and huge savings. I'm using Exergi LEDs to replace 50W halogen spots. The Exergis consume a measly 3.6W and give out almost as much light. That's a massive saving. The quality of LED lights is improving daily.
>
> I test-drove the Nissan Leaf (Nissan's all-electric vehicle) last year and I loved it. For many people it won't work as a primary vehicle, but for families who are looking for a second car, it excels. It is a fantastic drive, with range being the only limiting factor. Cost of motoring

with a Leaf is greatly reduced and as your utility adds more renewables to their generation set, your car becomes more and more Green!

My iPhone. This is a funny choice, I know, but before the iPhone I used to go through Nokias at a rate of a new phone every six to nine months as new features were launched. However, with the iPhone, new features are rolled out free with the latest download of its operating system—iOS. Less need now to shell out for new hardware; just download the latest free software update. The lifetime of my phones now is on the order of two years.

A Tough Crowd

There is a reason why Raftery calls his iPhone choice "a funny one." In many sustainability forums, Apple is the devil personified. It does not help that the company rejected in 2010 a resolution that would require it to publish a CSR (Corporate Social Responsibility) report centered on its greenhouse gas emissions, toxic waste, and recycling. Apple's "Board of Directors recommended shareholders vote against the resolution because they believe Apple has addressed sustainability reporting and that a formal report would add little value and involve unnecessary time and expense."[5]

Jeff Swartz, CEO of Timberland, the shoe company, has commented on his blog,

> CEOs of publicly traded companies in the fashion industry don't get the "pass" that comes to the supercool Apple leaders and their uber-cool company. Meaning, my shareholders and my consumers insist that we create profit, quarter by quarter, and that we do it . . . in a sustainable fashion, both in terms of environmental practice and in terms of transparency and safe working conditions in the supply chain. Why does a bootmaker get held to a higher standard than an iPad maker?[6]

Apple gets savaged for the workforce performance of Foxconn, its contract manufacturer in China. Multiple employee suicides were followed by an article in the London *Daily Mail* titled "You are NOT allowed to commit suicide: Workers in Chinese iPad factories forced to sign pledges,"[7] which sounded callous and made for very bad press. Soon after, combustible dust was blamed for a deadly explosion at a Foxconn plant.[8]

It does not matter that Apple points out its Supplier Code of Conduct draws on internationally recognized standards and "outlines expectations covering labor and human rights, health and safety, the environment, ethics, and management commitment. Apple monitors compliance with the Code through a rigorous program of onsite factory audits, followed by corrective action plans and verification measures."[9]

In its 2011 Supplier Responsibility Progress Report, Apple provided details such as:

> We expanded our training initiative beyond our final assembly manufacturers so that more workers in our supply base understand their rights and protections under local law and Apple's Code. Since launching in 2008, Apple's programs have trained more than 300,000 workers.
>
> We dedicated additional resources to protecting the rights of workers who move from their home country to work in factories in another country. Many of these immigrants are charged exorbitant fees that drive them into debt, an industrywide problem that Apple discovered in 2008 and that we classify as involuntary labor. In 2010, we continued our search for these violations, auditing all of our production suppliers in Taiwan and many in Malaysia and Singapore. As a result of Apple's audits and rigorous standards, foreign workers have been reimbursed $3.4 million in recruitment fee overcharges since 2008. We also trained suppliers on how to improve their recruiting practices, as well as on their legal and ethical obligations to foreign workers.
>
> We worked aggressively to prevent the hiring of underage workers. We equipped facilities with stronger age-verification tools, educated them on managing third-party recruiters, and held them accountable for the recruiting practices of affiliated schools and labor agencies. In addition, we are leading the industry by requiring suppliers to return underage workers to school and to finance their education.

It does not matter that Foxconn is also a key supplier for other tech vendors or that its Microsoft XBox 360 team threatened mass suicide in January 2012. It does not matter that Apple's retail stores and the App Store give opportunities to employees and entrepreneurs in many countries. For all the admiration and praise Apple gets in many other areas, it is fighting an uphill battle when it comes to the stainability image.

There are others who savage HP. You see that on so many emails these days: "think about the environment before printing." HP, of course, happens to be the largest printer company in the world.[10] Then there are others who would like HP to offer more soy- rather than petroleum-based inks.

Heather Clancy at *ZDNet* picks on another angle:

> HP's emissions related to employee travel were way up: a 49 percent increase related to air travel alone . . . the revelation is disappointing nonetheless, especially since Hewlett-Packard actually sells a line of telepresence technologies.[11]

Raftery comments again:

> Neither of these businesses (printing and devices) is particularly environmentally friendly and yet HP's founders spoke of HP's commitment to the environment as far back as 1957 in HP's first statement of corporate objectives, The HP Way.[12]

Blood Diamonds, Conflict Minerals, and Rare Earths

Hollywood introduced us to the moral issues in *Blood Diamond*, a movie starring Leonardo DiCaprio. That movie was set in Sierra Leone, but the action has since moved east in Africa, particularly to the Democratic Republic of the Congo, around a number of minerals we need in our high-tech devices.

The bland industry term for that is "conflict minerals," but as the *Huffington Post* reported:

> It's a war which most people know nothing about, despite the fact that we're all directly connected to it. Armed groups are fighting over the lucrative minerals that power our cell phones and laptops, leaving a trail of human destruction that has no equal globally since World War II.[13]

The UN General Assembly Mission Council is the mission and ministry agency of the Presbyterian Church, and on its blog it urges:

> The need and momentum is growing for a credible international certification system to ensure that minerals in consumer electronics and other products are not fueling rape and violence in eastern Congo. Ten years ago, a certification system addressing the trade in blood diamonds helped end wars in Sierra Leone and other West African countries. Similar systems have also led to significant progress in guaranteeing social and environmental standards through fair labor, forestry, and oil revenue transparency.[14]

Apple in the same report above highlighted what it is doing regarding "conflict minerals":

> We mapped the use of potential conflict minerals in our supply chain. We identified 142 Apple suppliers that use tantalum, tin, tungsten, or gold to manufacture components for Apple products and the 109 smelters they source from. Apple is also at the forefront of a joint effort with the EICC and the Global e-Sustainability Initiative (GeSI) that will help our suppliers source conflict-free materials.

Then there are rare earths, critical in many of today's technology products. iPods, for instance, contain small quantities of the rare earths dysprosium, neodymium, praseodymium, samarium, and terbium. Fiber-optic cables need erbium, europium, terbium, and yttrium.[15] Today, China is the major supplier of rare earths, but its own internal consumption has forced it to cut back on its exports. There are rumors of total Chinese bans on exports of certain of those minerals.

The U.S. Department of Energy published its Critical Materials Strategy analysis in December 2010. It identified five rare earth elements at "critical" risk of supply disruption within the next five years: europium, neodymium, terbium, yttrium, and dysprosium. [16]

The Mountain Pass Mine in the Mojave Desert in California has plentiful deposits of bastanite, from which several rare earths are extracted. Mining operations ceased at Mountain Pass in 2002 amid environmental concerns, although processing of previously mined ore continued at the site. Mining has since restarted to the consternation of environmentalists. The dilemma is minerals critical for newer cleantech versus environmental risks for which the mine was closed in the first place. And are we replacing one dwindling resource, fossil fuels, with another, rare earths? Is that really sustainability?

Then there is the promise of undersea mining. *New Scientist* editorialized:

> Now is the time to put in place legal frameworks to ensure that any rush for minerals will avoid the destructive effects of California's gold rush in the 19th century. The International Seabed Authority, which regulates mining claims in international waters, must adopt clear guidelines on conserving vent ecosystems.[17]

Packaging and Extended Supply Chains

An area where some tech companies have shown sustainability creativity is product packaging.

Dell claims that it slashed over 18 million pounds of packaging from 2008 to 2010 by focusing on a "three C's" packaging strategy: cube, content, and curbside recyclability.[18] The cube involves the reduction of size so that packaging is more efficient, from the size of the boxes that hold a product's components to the number of items that can be moved per shipping pallet. Dell has also boosted the percentage of recycled products that goes into its packaging content. The company increased its usage of recycled foam, and has used more recycled plastic as well. An estimated 9.5 million half-gallon milk jugs went into Dell's packaging—enough to stretch about 1,500 miles. Bamboo, a fast-growing, sustainable plant material, has also made its way into Dell's packing materials. The third is focus on materials customers can easily recycle in their neighborhood garbage collections.

Cisco described to Greenbuzz.com a pilot program that showed $24 million in annual savings.[19] Some of the techniques demonstrated included the reduction in size of antistatic and documentation bags and boxes, cutting plastic needs by 50 percent. Paper-based documentation was digitized and migrated to CDs or "pointer cards," 3 × 5-inch cards with online references. Nine 75-inch telepresence carton units now fit on a single truck, while only two units could be transported prior to re-engineering the packaging. Cisco also used recycled cushioning in one router family. Previously these cushions were largely made from virgin oil.

What gets measured gets managed. Using that thinking, Sprint, the U.S. telecom company, working with Trucost, a firm that helps assess emissions across a company's supply chain, released results of their analysis. It showed that Sprint's supplier emissions totaled 2.08M metric tons of CO_2, slightly more than Sprint's total direct and indirect emissions of 1.95M metric tons in 2009.[20]

IBM has asked its 28,000 suppliers in more than 90 countries to install management systems to gather data on their energy use, greenhouse gas emissions, and waste and recycling.[21]

The Changing Definition of Sustainability

Jeremiah Stone, in the SAP Labs division of the software company, provides a perspective on how the definition of sustainability has evolved in just the last couple of years.

In Stone's words:

> When we launched our program in late 2008, SAP's starting position was a strong solution set in environment, health, and safety management with its long-time partner, Technidata. Otherwise, it was a blank sheet as far as our sustainability portfolio was concerned.
>
> First, we predicted a strong shift from sustainability reporting to performance management solutions, particularly around greenhouse gas (GHG) emissions and reduction in energy consumption. Second, we expected that this increased focus on GHG emissions and corresponding energy costs would drive significant investment in carbon footprint and energy management solutions. Finally, we expected to see an accelerated demand for solutions addressing Design for Environment (DFE), and sourcing driven by stricter procurement policies.
>
> Two major factors have delayed market maturation following our initial investment. The first was the "great recession" of 2009, and second was the failure of COP-15 (the 2009 Copenhagen Climate Convention) to deliver a global carbon reduction framework. These two factors have led to a slowed proliferation of Chief Sustainability Officers (CSO) with significant budgets for IT purchases or projects, and a cautious carbon legislation climate particularly in the United States and Australia.

In contrast, the BP Macondo well blowout, several mining disasters, and multiple food contamination scares from China to Germany have increased short-term awareness of operational risk and product compliance and stewardship. Further, the continued growth of the global consumer, or "middle" class continues its march unimpeded, driving demand for commodities ranging from rare earth minerals required for production of high tech gadgets to increased energy required to feed these gadgets and increasing proliferation of modern housing and facilities.

These high profile events and the energy efficiency megatrend are driving increased demand for:

Predictive risk management—Data drawn from core systems of record like human resources, asset and supplier management are driving predictive analytics to identify emergent risk in advance of traditional expert-based systems. Customers here span high-risk industries such as mining and mills, oil and gas (up- and downstream), chemicals, utilities, transportation and logistics, industrial machinery, and component manufacturing.

Supply chain traceability—This helps track movements of product at a lot or batch level across the extended supply chain to support market withdrawals or recalls on a finely targeted basis. Customers here span manufacturing, with added emphasis from food and industrial machinery and industries with expensive components.

Energy and environmental resource management—Tools for energy efficiency, optimized energy procurement, and emissions management. Industry focus here mirrors that of the predictive risk management opportunity, with the additional presence of banking and retail due to their large real estate portfolios that need building energy management.

Enterprise environmental accounting—Ability to account at a very fine level of detail the environmental "cost" of product based upon pathway analysis and emission data in stark contrast to standard Lifecycle Cost Assessment methodologies. Demand is primarily coming from consumer products customers who are seeking to drive accountability and performance against environmental goals. The core driver here is brand differentiation in segments where energy, water, carbon, etc., costs are meaningful for end-consumers.

So, we are seeing more and more examples of customers embedding SAP Sustainability solutions into their businesses and means of production. This bodes well for a continued transition of Sustainability from an aspiration to a core requirement of business success.

Ray Lane, lead cleantech partner at Kleiner Perkins, the venture capital firm, provides his perspective on the changing definition:

> You can see the evolution in the decade Kleiner Perkins has been investing in cleantech. We started out with biofuels, solar and wind. We next invested in conversion tech, coal to gas, thermal electrics, and waste heat to energy. Now we're investing in storage, fuel cells, etc., and in water. Two years ago we hadn't done anything in water, and now we've done three investments in clean water. We are also focused on agriculture from productivity of seeds to producing sugars.

The changing expectations of the elite around sustainability are a good segue to Part III (after the Google case study) where we cover changing regulatory and societal expectations of technology.

Conclusion

To be considered a technology elite, it is increasingly expected that you put sustainability high on your self-evaluation scorecard. The definition of sustainability, however, gets more ambitious by the day. Let's next look at Google's wide array of Green initiatives.

Case Study: Google's Green Initiatives

Jim Miller has an impressive resume. His career includes stints at:

> Intel, at the birth of the Pentium; Amazon.com, in the early stages of e-commerce; Cisco, when broadband exploded; First Solar, as part of the green/solar resurgence; and now Google, where as Vice President for Worldwide Operations he is in the engine room for the emergence of cloud computing.[22]

Most of his employers prior to Google had hardware/logistics elements. So as he was being recruited by Google, he wondered how different it would be to work for a software company for a change. Even after months of due diligence on both sides, "My job offer letter had so few details about the operations I would be running, that I had to take a leap of faith in joining Google," says Miller. That is no surprise, since Google is extremely secretive about its global operations.

Of course, this "software" company has lots of physical assets in its data centers, self-driving cars, and leased satellites that Miller is admirably qualified to optimize. And they are at a scale that challenges even a rocket scientist like Miller.

On any given day approximately 10 percent of the world's Internet traffic traverses the Google infrastructure. That comes from more than 1 billion query searches per day, over 500,000 Android activations a day, 2 billion YouTube videos watched each day and 120 million Chrome browser users. That does not even begin to account for demands from applications like Google Maps, Earth, Voice, Translate, Mail, Docs, Google+, and countless others. Add to that the logistical and energy needs of a global employee base. As Miller says, "My operations team is small (175 out of total employee count of 25,000) but it is intense and we spend a substantial portion of Google dollars (2010 revenues of $29 billion)."

Just don't ask him to publicly discuss how much he spends.

The operations focus is on cost, systems availability, efficiency, and being "green." Google has a stated goal of being carbon neutral as a company. Its accomplishments to date are impressive. Its data centers claim to use about half the energy of a typical data center.

Let's explore a few dimensions of the "green focus" at Google.

Google Energy LLC

Google has a subsidiary certified by the Federal Energy Regulatory Commission (FERC) to purchase power and resell it to wholesale customers. While the initial market reaction was Enron-esque and that it was designed to speculate around energy markets, Google presents it as facilitating purchases of renewable energy.

Google explains:

> The plain truth is that the electric grid, with its mix of renewable and fossil generation, is an extremely useful and important tool for a data center operator, and with current technologies, renewable energy alone is not sufficiently reliable to power a data center.[23]

So, Google buys electricity directly from a renewable project developer in the form of a power purchase agreement, or PPA. Their first PPA was with NextEra Energy Resources. Google agreed to buy 114 MW of wind power for 20 years from a project in Ames, Iowa, directed to a data center in Council Bluffs, Iowa. In Oklahoma, they added just over 100 MW of wind power for the data center in Mayes County. Since then, Google has announced another commitment of $38.8 million with NextEra in North Dakota.

This is where the energy reselling is involved. Google sells the power acquired under the PPAs back to the grid at the local, wholesale price. Today, because generic "grid" power is cheaper than renewable power, this may result in a slight net loss for Google, but longer-term, Google is betting the economics will reverse. In the process of selling, Google strips the renewable energy credits (RECs) to apply in the next step.

The data centers at Iowa and Oklahoma will be largely powered by conventional power from the local grid. Since the RECs were produced on the same grid, one REC represents one MWh of renewable power "used" at their data centers, displacing one MWh of local conventional power, and the data centers are considered carbon free.

While the accounting makes your head spin, the important nugget is that Google has given NextEra a firm, long-term commitment to build wind farms. Mike O'Sullivan, senior vice president of development for NextEra, was quoted as saying, "With the support of customers like Google Energy, we've built our wind fleet from fewer than 500 megawatts a decade ago to nearly 8,300 megawatts—the largest fleet in North America today."[24]

It also allows Google to decouple the data center location decision from the renewable energy source. Data center location decisions are driven by many factors including latency, taxation, talent, and real estate. As Google says, "Building a data center in an optimal area for renewable development would result in increased latency for our users and the inefficient use of land better used for renewable energy."

Investment in Other Renewables

Google has been investing in various start-ups to gain access to other forms of renewable energy. BrightSource and eSolar focus on concentrated solar energy and use swiveling mirrors to reflect sunlight to heat towers of water. The resulting steam is used to generate electricity.

The investments in AltaRock Energy and Potter Drilling were to get access to enhanced geothermal energy. The principle is to drill deep enough to get to the hot core of the earth, then pump water into it and use the resulting steam to create energy. Think of them as manmade geysers.

Google also invested in a company called Makani Power, which is leveraging high-altitude wind. One of their concepts is to fly kites with propellers. As the propellers spin they act like turbines, and the power is circled down a cable back to the ground.

In 2011, Google allocated $160 million of funds for the Alta Wind Energy Center (AWEC) in Tehachapi, California. It also allocated $280 million toward residential solar power; $100 million of that went toward a majority stake in the 845-megawatt Shepherds Flat wind power project in Oregon. Another $168 million investment went into BrightSource Energy's 2,600 megawatts Ivanpah Solar Electric Generating System in California's Mojave Desert.[25]

Google has also invested in the development stage of the "Atlantic Wind Connection"—an underwater transmission network that can harvest electricity from wind farms off the Mid-Atlantic coast and could, when finished by 2020, power 1.9 million homes across a 350-mile network across Virginia, New York, and New Jersey.

All told, Google's investment in such renewable projects now exceeds over $1 billion.

In addition to renewables, Google has made investments (though relatively small ones) in cleantech startups, including battery maker ActaCell, electric vehicle maker Aptera, efficient car maker Next Autoworks, neighbor-to-neighbor car sharing company RelayRides, weather insurance company WeatherBill, smart grid company Silver Spring Networks, biofuel maker Cool Planet Biofuels, and efficient power gear conversion startup Transphorm.[26]

Ray Lane, a lead cleantech partner at the venture capital firm Kleiner Perkins, says: "We think the world of Google Ventures. In the hypercompetitive world of venture investing, we share intelligence, co-invest, and do a number of collaborative activities. Google, as a customer, has also been an early adopter of products like fuel cell technology of our portfolio company, Bloom Energy. No question, Google bleeds green."

Data Center Efficiency

In 2010, Google took over a former paper mill in Hamina, Finland, and retrofitted it into a data center. It continues to use a seawater tunnel that was built for the paper mill in the 1950s. The seawater passes through four different straining systems. This reduces corrosion from the salt and other minerals in the seawater before it reaches the heat exchanger and is used to cool the data center. On the way out, water then moves to a tempering building, where it mixes with a separate source of the seawater, so it is cooled before returning to the Gulf of Finland. The goal is to "return to a temperature that is much more similar to the inlet temperature, so as to minimize environmental impact in this area."[27]

Google's experience at Hamina and with every new data center it opens around the world adds to a sizable bag of tricks from a decade of running data centers. It increasingly shares with the world some of the best practices it has accumulated.

Its practices include:

- Measure PUE: "We use a ratio called PUE—Power Usage Effectiveness—to help us reduce energy used for noncomputing, like cooling and power distribution. To effectively use PUE, it's important to measure often—we sample at least once per second. It's even more important to capture energy data over the entire year—seasonal weather variations have a notable effect on PUE."
- Manage air flow: "Thermal modeling using computational fluid dynamics (CFD) can help you quickly characterize and optimize air flow for your facility without many disruptive reorganizations of your computing room."
- Adjust the thermostat: Raising the cold aisle temperature will reduce facility energy use. Don't try to run your cold aisle at 70°F; set the temperature at 80°F or higher—virtually all equipment manufacturers allow this.
- Use "free cooling": Since chillers are the dominant energy-using component of the cooling infrastructure, minimizing their use is typically the largest opportunity for savings. There is no one "right" way to free cool—but water or air-side economizers are proven and readily available.
- Optimize power distribution: Minimize power distribution losses by eliminating as many power conversion steps as possible. One of the largest losses in data center power distribution is from the uninterruptible power supply (UPS); so be sure to specify a high-efficiency model. Also keep as high a voltage as close to the load as feasible to reduce line losses. We also recommend using energy efficient IT equipment, especially those with high efficiency power supplies. Look for the EnergyStar label for future server purchases.[28]

All these lead to an impressive Google statement: "In the time it takes to do a Google search, your own personal computer will likely use more energy than we will use to answer your query."

RE<C

Through November 2011, Google.Org, one of Google's philanthropic initiatives, funded a project called RE<C—to develop one gigawatt of renewable energy capacity (enough to power a city the size of San Francisco) at a price cheaper than coal, in years, not decades.

Google's former "green energy czar" Bill Weihl explained:

> We've learned that a small team of smart people with basic technical expertise and the freedom to really innovate can do something quite remarkable, and we wanted to see if that really could be true for alternative energies. One of the keys there is the freedom to go after a really aggressive goal, and so we set a goal of making renewable energy cheaper than coal—it's a very simple, kind of audacious and crazy goal.[29]

Some of the R&D projects included an effort to design and build low-cost heliostats, mirrors that track the sun, and reflect sunlight to concentrate solar energy. Another worked on a solar Brayton engine (a gas turbine engine like those currently used in jet aircraft, but powered by sunlight) that would heat air to drive a turbine and generate electricity.

Consumer and Other Applications

Google has created a $280 million fund with home solar installer SolarCity. This is Google's largest investment in clean power to date, and its first in-home rooftop solar.

For several years Google offered an app called PowerMeter, a free energy monitoring tool that helped consumers save energy and money. Using energy information provided by utility smart meters and energy monitoring devices, Google PowerMeter enabled you to view your home's energy consumption from anywhere online.

Google has also built the largest corporate installation of solar panels at headquarters campus in Mountain View, CA. Over 9,000 solar panels means that the "installed capacity of this solar grid is 1.6MW.... In one day the system generated 9,468 kilowatt-hours of electricity. This is enough electricity to power 83,000 hours of flat-screen TV viewing each day."[30] Google was an early adopter of Bloom Energy's boxes on its campus. "Over the first 18 months the project has had 98 percent availability and delivered 3.8 million kWh of electricity."[31]

Then there are the other less prominent sustainability moves on its main campus. Robyn Beavers of Google provides some examples:

> When we're done with (our carpet), we can send it back to the manufacturer and they grind it up into little pellets and use it again in the supply stream so it never ends up in a landfill. In some of our window shades and the textiles we use in our cubicles, we focus on eliminating toxins. We have filtered water everywhere, we have 90 percent fresh air coming into the building throughout the day—a lot of stuff you can't really see.[32]

Summarizes Weihl:

> I believe that the problems we're facing are solvable, but they're not going to solve themselves. And solving them is either going to require spending a lot more money on energy than we're spending today, which I think is probably a non-starter, or it's going to require major technological innovation. That's where I think Google can help.

Part III

Outside Influences on the Technology Elite

Chapter 18

Making Regulators More Tech-Elite

During the BP Gulf of Mexico oil spill in 2010, U.S. Secretary of Energy Steven Chu blogged, "My job has been to oversee the federal science team—a group of top scientists from the Department of Energy's national labs, the federal government, and academia, along with outside industry experts. . . ."[1]

Around the same time, as the U.S. Department of Transportation was investigating multiple occurrences of sudden acceleration in Toyota vehicles, it announced it had brought in NASA engineers to help. The NASA charter was to determine "if there are design and implementation vulnerabilities in the Toyota Electronic Throttle Control System Intelligent (ETCS-i) that could cause UAs (unintended accelerations) and whether those vulnerabilities, if substantiated, could realistically occur in consumers' use of these vehicles."

The NASA team reported after its analysis:

> Because proof that the ETCS-i caused the reported UAs was not found does not mean it could not occur. However, the testing and analysis described in this report did not find that TMC ETCS-i electronics are a likely cause of large throttle openings as described in the VOQs (Vehicle Owners' Questionnaire).[2]

In an interview, Secretary Chu explained the reason for the large brain trust during the BP spill:

> After the [Space Shuttle] Challenger accident, the U.S. government formed a panel of very, very bright scientists and engineers to come together and figure out what happened and what could be done in the future to prevent it. Most of the people on that panel were not aeronautics experts, not rocket experts or NASA experts. They were very smart people who had a broad range of knowledge and experience. This is actually what you want: you want a set of fresh eyes, people who can propose potential out-of-the-box solutions, who might foresee what might go wrong. If you're an expert and you're used to certain things done certain ways, that limits your ability to cast a wider net, and so one of the most important things that we're doing at the national laboratories is putting together these scientific teams, many of whom would be considered nonexperts. In times like this, those are many of the people you want.[3]

Rocket scientists helping on auto investigations, nuclear physicists helping on ocean-based investigations, and a multitude of other specialists helping with rocket science. We saw glimpses of government efficiency and innovation in earlier chapters with the examples of the country of Estonia, the Hillsborough County Tax Collectors Office, and Roosevelt Island. Overall, though, it is becoming clear that technology is stretching the capabilities of regulators. The range of technical skills we need in our regulators becomes very apparent when you look at the 3M Periodic Table we present in the case study with its 46 "technology platforms" from Biotechnology to Optical Communications.

Regulatory Challenges Galore

If there is a poster child for an agency that is overwhelmed by technology advances, it is the U.S. Patent Office. Even though a quarter of patent applications come from California, it does not have a branch office there. The Office is so backed up that applications don't get looked at for years. It is not uncommon for patents to take three to five years to get issued. Until recently, the Office, which should be tech savvy itself, would not accept digital applications.

Then, as we saw in Chapter 14, interpretation of those patents is even more convoluted, leading to many legal battles. Google's chief lawyer Kent Walker has been quoted as saying the smartphone industry is using patents in an arms race that is "gumming up the works of innovation."[4]

Indeed, one of the major reasons Google bid $12.5 billion to acquire Motorola Mobility was its vault of 17,000 patents and another 7,500 in progress.

Commented Rafe Needleman: "The accumulation of patent portfolios into a smaller number of bigger players, which themselves are locked in a deadly standoff, has the real potential to slow down the pace of innovation. Which is precisely the opposite reason the patent system was created."[5]

It is estimated that over 200,000 emergency 911 calls are placed daily in the United States via mobile phones, and even though the phones can relay GPS locations, the 911 centers have to ask for specific location information, when the caller is likely under extreme stress. Additionally, VoIP lines have been replacing landlines in countless homes and businesses for over a decade now. Yet, it's only in 2011 that the FCC has announced plans for Next-Generation 911, "seamless, end-to-end IP-based communication of emergency-related voice, text, data, photos, and video between the public and public safety answering points."

There are many examples of inadequate regulation in the financial sector. *The New Yorker* summarized:

> Financial regulators let A.I.G. write more than half a trillion dollars of credit-default protection without making a noise. The S.E.C. failed to spot the frauds at Enron and WorldCom, gave Bernie Madoff a clean bill of health, and decided to let Wall Street investment banks take on

obscene amounts of leverage, while other regulators ignored myriad signs of fraud and recklessness in the subprime-mortgage market.[6]

Frank Scavo, who runs Computer Economics, says,

We've been benchmarking IT organizations since 1990, so we have quite a bit of information on historical IT spending trends. Based on our data, we know that financial services organizations are among the most IT-intensive industries, no matter how you measure it. For example, this year, the typical commercial bank is spending 6 to 7 percent of its revenue on IT. In contrast, the typical manufacturing company only spends 1 to 2 percent of its revenues on IT. This is because banks, insurance companies, and other financial institutions are really information businesses. On the surface, they are developing, marketing, and selling financial services. At their core, their business is all about leveraging information for financial gain.

Do the regulatory authorities have the skills and systems to oversee these institutions? After the last financial crisis, you have to wonder.

Life sciences are another sector that is very IT-intensive. According to our research, the typical life sciences organization today spends 4 to 5 percent of its revenue on IT, and this doesn't include all the technology investments taking place outside of the traditional IT organization. For example, most high-end medical devices, such as diagnostic imaging and radiation therapy equipment, are really smart devices in every sense. They process large amounts of information and are increasingly being connected to hospital networks. At the other end of the spectrum, small start-ups are figuring out how to use smartphones to displace low-end medical devices, such as stethoscopes, blood pressure monitors, and glucose testers. You can even buy apps like these on iTunes for a few dollars.

Do regulatory agencies such as the U.S. Food and Drug Administration (FDA) have the resources to ensure such devices are safe and effective? To the FDA's credit, it has indicated that it does not want to stand in the way of technology adoption. But whether it has adequate technical staff to review these new products coming to market is another question.

Mark Cuban, described in earlier chapters, is much more direct about what financial regulators should be doing:

> [They] have got to start to recognize that traders are not investors and vice versa and treat them differently. Different regulations. Different tax structure. Different oversight.[7]

Not Just U.S. Regulators

The head of the International Atomic Energy Agency (IAEA) has acknowledged that he would like to see his agency more involved in damage control from any future nuclear disaster. These were comments prompted by criticism of the IAEA's role in the Fukushima accident after the Japanese earthquake and tsunami in early 2011. "The IAEA itself will acknowledge privately that it did not cover itself in glory," says James Acton, who studies nuclear policy at the Carnegie Endowment for International Peace in Washington, DC.[8]

The Air France 447 crash off the Brazilian coast in 2009 raised a number of regulatory issues. The icing of Thales AA pitot tubes, which help calculate airspeed, had been shown to be a regular problem and yet "Regulators simply asked Airbus to watch the problem and report back in a year." In our days of streaming TV and music, why do planes still store critical data on old technology called black boxes? In the case of the Air France flight, it took over two years to recover the black box from the bottom of the ocean. As a *New York Times* article suggested, we should be aiming for at least partial data streaming directly from the plane during events like failure of the autopilot.[9]

In July 2011, a report by the UK Public Administration Select Committee (PASC) of the UK House of Commons said, "The Government's over-reliance on large contractors for its IT needs combined with a lack of in-house skills is a 'recipe for rip-offs.' The committee found that as a result IT procurement too often resulted in late, over-budget IT systems that are not fit for purpose."[10]

The Shifting Winds

Robert Hoffman has more than two decades of policymaking experience in Washington, including 11 years as a legislative aide and director in the

U.S. Senate. He also has more than a decade of experience as a public policy manager and advocate for Oracle Corporation and Cognizant Technology Solutions.

He summarizes trends in technology oversight in Washington:

> The U.S. federal government has long struggled with regulating information technology. U.S. state governments and the European Union have become comfortable playing the role of IT consumer advocate, and strictly defining the responsibilities and requirements of IT developers, vendors, and users on how sensitive personal information is stored. For nearly two decades, Washington has hesitated diving so confidently into the IT regulatory pool.
>
> Even when the horror of 9/11 brought even more compelling arguments for tighter government regulation of crypto-products and IT systems, Washington again hesitated. It wasn't just the threat of a mass exodus of high-technology and high-paying jobs that prompted Washington to hesitate. Legislators and regulators did not have a firm grasp of the technology landscape itself. Sending emails or surfing websites constituted the extent of a legislator's or regulator's exposure to technology. Indeed, the most tech-savvy people on Capitol Hill in the 1990s were overwhelmingly the young twentysomethings that wired-up the fledgling client-server operations and programmed the first mobile phones in each congressional office. If there were dominant regulatory arenas that the IT industry had to confront in Washington over the past two decades, they were antitrust and export controls. After all, policymakers may not fully understand technology itself, but they could easily conclude that too much of something in the hands of one or a few, whether that something was soft drinks or software, can't be good for the U.S. economy. Similarly, they understood state-of-the-art technology may be good for financial institutions, but not foreign terrorists.

Hoffman continues with a focus on today:

> So, fast-forward to 2011, and let's review the key policy and regulatory issues that are on the IT policy agenda: cyber-security, data privacy, data breach reporting requirements, standards for critical infrastructure protection, and Internet neutrality. Today, there is no one federal department or agency that has singular regulatory authority over IT,

and those that could assert such authority won't do so without clear congressional authorization.

For DC veterans, this agenda has a familiar ring. Indeed, almost all of today's issues and regulatory challenges were on the agenda in 2001. The technologies certainly have advanced, but the issues have marked time and could not be resolved. This is due to many of the same reasons that influenced the encryption debate: the fear of stifling innovation, an uncertainty about the underlying technology itself, and competing points of view from within the sector, which suggested that policy solutions would effectively distort the market by picking winners and losers. When it comes to managing IT policy, it's been Groundhog Day for well over a decade in Washington.

Hoffman on likely changes:

I don't see that lasting for much longer. For the next two to three years, regulators and policymakers are likely to achieve more regulatory authority and policy certainty over the IT industry than ever before. True, given the past track record, that won't be hard, but several fundamental factors are working to create an environment where legislators and regulators will look at IT policy with greater creativity and confidence. Those factors are:

- Tech-savvy policymakers are coming of age. The twentysomethings that first connected Capitol Hill to the information superhighway are now in positions of authority—on the Hill, in regulatory bodies, in trade associations, and in industry. They are showing nimbleness in response to the emerging trends in IT, ranging from mobility to social media to cloud computing. Yes, the IT industry is still a young person's sector, but there isn't the kind of generation/knowledge gap between IT and DC that we witnessed in the 1990s.
- The federal government is becoming not just a major IT user, but a better IT manager. At roughly $80 billion per year, the federal government is the single largest buyer of technology in the United States, but only in recent years has it decided to manage its IT infrastructure in ways that the private sector has done for years. It appointed its first-ever CIO, who pursued much-needed reforms to consolidate servers, and provide greater transparency in IT management. The challenges and uncertainties public agencies

face about managing IT systems in a cloud environment are the same as those faced by private sector firms.

- As the federal agencies become better IT managers, they are certain to change the dynamic of the federal government's relationship with the private sector. Standards to protect critical infrastructures, methods to respond to emerging cyber threats, and data privacy are clearly areas of mutual concern to private and public sectors. That would suggest that a more collaborative public IT manager-to-private IT manager approach to respond to these issues would be a compelling option for governance, rather than a traditional, mandate-driven public regulator-private regulatee approach.

Hoffman summarizes:

> One of the challenges for the federal government, as its agencies continue to better manage IT infrastructures and be responsive to new trends, is whether it can reconcile and integrate these dual roles of manager and regulator of IT. Are the agency regulators talking to the agency's CIOs, CSOs and CPOs, and vice versa? Such intra- and interagency collaboration is likely to make the federal government better at both managing and regulating IT, while staying true to the long-standing principle of promoting IT innovation and doing no harm.

Conclusion

The regulation of technology is going through significant change. This is causing the technology elite to evolve how they interact with regulators and other market watchers. The 3M "Periodic Table," described in the following case study, is what market watchers will increasingly have to be intimately familiar with.

Case Study: 3M's "Periodic Table"

3M, a company founded in 1902, is a remarkably diversified entity. David Meline, its CFO, broke out the $ 27 billion in 2010 revenues at an investor conference in June 2011. They were distributed across industry sectors: 32 percent from Industrial and Transportation sectors, 17 percent from Healthcare, 14 percent from Consumer and Office, 14 percent from Displays and Graphics, 12 percent from Safety, Security, and Protection Services, and 11 percent from Electronics and Communications.

Geographically, it got 35 percent of revenues from the United States, 23 percent from Europe, 31 percent from Asia Pacific, and 11 percent from Latin America.

As Meline got to slide 14, he presented a table similar to one we studied in Chemistry class. But instead of H for hydrogen and Au for Gold it showed 46 3M "technology platforms"—Bi for Biotechnology, Op for Optical Communications, and so on. 3M innovates by finding "uncommon connections" of these platforms to create unique solutions for customers across all six of their focus industry sectors shown above.

From slide 15 on, he was back presenting financial numbers.

That one slide though, drove home 3M's amazing technology diversity. 3M is fundamentally a science-based company. With more than 55,000 products, 3M continues to demonstrate an uncanny ability to combine highly innovative technologies in new and unexpected ways.

It is a remarkable turnaround for a company with a "star-crossed" beginning as described in its history published on its 100th birthday. It detailed:

> First, a worthless mineral, then virtually no sales, poor product quality, and formidable competition. All the founders had to keep them going was perseverance, a spirit of survival and optimism. What would happen next? It was the equivalent of the sky falling, only at ground level. 3M built its new plant, a two-story, 85-foot-by-165-foot structure with a basement. It wasn't the best construction, but it was all the budget allowed. When raw materials arrived from Duluth and were stacked on the first floor, one Saturday, the weight tested the timbers—and the timbers lost. The floor of the new plant collapsed and every carton, bag, and container landed in a heap in the basement.

Table 18.1 3M's "Periodic Table"

Symbol	Platform	Description
Ac	Acoustic Control	Loud, unpleasant noises aren't just irritating—they are also a threat to our health and safety. 3M has engineered select nonwovens, viscoelastic polymers, and other materials to absorb sound, dampen vibration, and control undesirable noise. In automobiles these materials create a quieter passenger compartment. In computer disk drives, they reduce irritating sounds and help increase access speed and storage capacity. In the workplace they reduce distraction, protect hearing, and allow communication.
Bi	Biotechnology	To create products for the health sciences, 3M applies its unusual combination of strengths in material science, surface characterization, filtration, and biology. Among the results: remarkably durable dental restoratives, biocompatible surgical tapes and dressings, transdermal patches for drug delivery, antimicrobial cleansers, and sterilization sensors.
Di	Display	3M display materials manipulate the transmission and reflection of light. By combining our expertise in multilayer films, lenses, and microreplication, we are able to increase the clarity, durability, or brightness of virtually all types of electronic displays—from liquid crystal displays in handheld devices, to LCD monitors and televisions, to gigantic projection displays. These materials are also used in privacy filters for laptop computers and for automated teller machines.
Em	Electronic Materials	Advances in electronic components—from semiconductors and interconnects to data storage and displays—often begin with breakthroughs in the organic, inorganic, and metallic materials from which these components are fabricated. 3M contributes to improvements in manufacturing and performance by applying its strengths in many of the other building blocks of today's electronic devices, including fluorochemicals, adhesives, abrasives, membranes, electrodes, and electrolytes.

Table 18.1 (*Continued*)

Symbol	Platform	Description
Im	Imaging	3M understands visual communication. We made the first thermofax machine for transmitting images. We redefined advertising with the first digitally printed, wall-sized graphics. We invented multilayer optical films that create crisp, vibrant images for electronic displays and counterfeit-resistant security marks. We've refined our inks and pigments to ensure true, bright colors that last, even under the harshest conditions. Building on decades of experience in print and digital technologies, 3M continues to introduce sophisticated films, inks, and software for print, electronic displays, and signage.
Md	Medical Data Management	In more than 6,000 hospitals around the world, 3M's advanced software and consulting services help organizations capture and classify healthcare data, allocate resources, comply with regulations, manage revenue, and ultimately improve the quality of patient care. Our solutions bring significant cost savings to healthcare organizations and help governments and health policy agencies worldwide manage costs and improve the outcomes of their local and national healthcare systems.
Mr	Microreplication	3M has pioneered the use of tiny, precisely shaped structures to give materials new physical, chemical, or optical properties. Microreplicated prisms are used in road signs, electronic displays, and exterior building illumination to capture and reflect light more brilliantly. Minute pyramids make 3M's structured abrasives work better and wear more evenly. And microreplicated channels can even direct fluids using capillary action, an application with enormous potential for biomedical products.
Op	Optical Communications	With the growth of online services such as Internet telephony and video-on-demand, the demand for greater bandwidth is escalating rapidly. 3M's fiber optic splices and connectors—based on our

(*Continued*)

Table 18.1 *(Continued)*

Symbol	Platform	Description
		expertise in materials science and precision processing—help service providers build and maintain networks to satisfy customer expectations for speed and reliability. Our strength in optoelectronics is also reflected in 3M display components, flexible electronics, and sensors for medical, security, and defense markets.
Rf	RFID	Radio frequency identification (RFID) has become an extremely reliable, valuable, and pervasive tool for monitoring the movement and status of items. 3M first developed this wireless tracking technology for use in library circulation systems. Today, our customers use it to keep tabs on pharmaceutical shipments, aircraft engine maintenance, tissue samples, legal and medical files, and much more. With constantly improving software and hardware, 3M RFID helps increase productivity while cutting the costs associated with theft and loss.
Se	Sensors	The faint flutter in a newborn's heart, a crucial shift in temperature that destroys a vaccine's potency, a genetic marker for cystic fibrosis—these and other indications can be detected and communicated by 3M sensors and other sensors relying on 3M components. Our technologies respond to a comprehensive array of stimuli, including heat, light, pressure, radio, chemical, biological, and other signals. Sensor applications are also comprehensive, ranging from low-cost indicators—such as single-use sterilization test strips—and advanced, automated diagnostic devices.

Source: 3M.

How does a company so diverse communicate with its customers and market watchers? One creative way has been to present its "Periodic Table" on its website, allowing viewers to drill into each platform.

Table 18.1 extracts 10 of those platforms, and you can see the breadth of markets it supplies technology products and components. This shows less than a quarter of its platforms.

Chapter 19

Society's Changing View of Technology

While this book is about all kinds of smart products and services and technologically sophisticated companies, we would be remiss to not point out we live in an uneven, "unelite" society. A relevant question to ask is whether society at large is ready for the massive coming tsunami of technology. This chapter presents some considerations that cause pause.

Our Fragmented Society

At one extreme is Fred Wilson, a venture capitalist who, with his wife Joanne, encourages their kids to be comfortable with all kinds of technology. "The parents and kids publish a combined nine blogs. They bring a duffle bag on family trips just to carry all the cords, adapters, and batteries for their electronic devices."[1]

And then at the other extreme there are what *USA Today* calls the "Tech-Nos," including folks like Joan Brady: "No, she doesn't e-mail. And, really, she does not need you to call her and read the latest e-mail joke to her. She knows what she's missing, and she's grateful for it every day."[2]

It is estimated that the number of U.S. mothers who have used midwives to deliver babies naturally has doubled over the past several decades. At least some insurance companies are starting to pay for alternative healthcare like acupuncture and chiropractic care—relatively low-tech services. (Of course, in a sign of the times, we now have laser acupuncture and expert systems to suggest precise acu-points to be needled depending on the ailment being treated.)

Then there are other customers that are cynical of technology—with good reason. Banks sold automated teller machines (ATMs) as customer self-service, and then tacked on fees for that self-service. Companies are now selling electronic invoices as "green" and progressive but then trying to tack on fees for that "privilege." Hollywood has made money on the same content in VHS, DVD, and now BluRay formats. Customers feel that their privacy is not protected and their lives are subject to surveillance as technology increases in products. These customers are not Luddites—just wary.

In the middle between Wilson and Brady is someone like Brian Sommer. Sommer is a former Accenture partner and is now founder of a technology consulting firm, Techventive, and a *ZDNet* blogger. He is not a technology Luddite by any means; he just has a cautious philosophy on technology adoption. He explains:

> I'm that fellow who repairs all the neighborhood computers and sets up everyone's wireless networks. Every desktop, laptop, and netbook I own, I've upgraded the memory, the disk drive, or some other component within them. I get technology and I've written, spoken, and researched the space for decades.
>
> I don't, however, own a smartphone and I don't have a Facebook page.
>
> I'm not making some sort of antisocial statement. Actually, it's just the opposite, as I may be a lot more social than many Facebook users. I just don't like the privacy tradeoffs many new technologies inflict on

the public. I also don't buy into the premise that posting content on a website is the same as really investing in a friendship.

There's a fascination in tech circles that one successful technology can be made better by smashing another technology into the former. Cell phones used to be about making telephone calls. Now, some cell phones are computers, GPS devices, cameras, bill payment tools and music players. Yes, that sounds convenient, but I leave a GPS in my car. My digital camera has a great zoom capability that is nonexistent on a smartphone. My iPod stays mostly in my briefcase. I have a netbook and laptops for computers and they are still more powerful and secure than my phone.

I like "Just Enough" technologies. They're simple, straightforward, and easily replaceable. If I lose or break my cell phone, I can get another for under $20. It's similar for laptops and cameras. Not only are my repair bills low (or nonexistent), I never suffer complete technology failure the way those who put everything into one device do. If a thief stole my phone, they're not getting much. What would they get if they stole your smartphone?

Just enough technology gives me peace of mind. I don't worry about my cell provider or anyone else knowing exactly where I am. My privacy is easier to protect when I don't have one of these über-devices tracking and controlling every aspect of my work and professional life. Just enough technology is really inexpensive, too. I can cheaply change out any component. And, just enough technology means I'm never really tied up with a single provider and I'm not on contract with any firm.

I'm not on Facebook either as I find it's easier and more rewarding to have friendships the old-fashioned way. I use a cell phone to talk to friends and family, not to update my social network page. I make an effort to be someone's friend, to listen to them and to discuss their needs, wants, and challenges. I invest in them. I don't invest my scarce time and attention to building website monuments to my egotism or narcissism. When I see someone at a rock concert updating their social network page during a number, I feel sorry for this person and their date. They may have gone to the event together but they aren't bonding and their friendship is straining.

I'm staying off Facebook and I'm keeping my old "just-enough" technologies for now. Yes, you can try to shame me into getting a smartphone or a Facebook account, but I'll probably resist it. I want to keep my privacy and the flexibility that permits low cost. Plus, it helps keep me grounded in what's important: my time, my family, and building enduring relationships. If you'd like to have a real friendship someday, give me or someone like me a real call and let's connect in a positive way.

Our Digital Generation Gap

Wired magazine listed 100 things our kids will likely go "huh?" about, including the following:

- Inserting a VHS tape into a VCR to watch a movie or to record something.
- Rotary dial televisions with no remote control.
- The scream of a modem connecting.
- The buzz of a dot-matrix printer.
- Booting your computer off of a floppy disk.
- Using a road atlas to get from A to B.
- Doing bank business only when the bank is open.
- Sending that film away to be processed.
- Vacuum cleaners with bags in them.
- Not knowing who was calling you on the phone.[3]

Asked about that list, Charlie Bess of HP goes, "And don't forget doing math in their head!"

Caitlin McNally is associate producer of the PBS show *Growing Up Digital,* which points out that "the Internet has created the greatest generation gap since rock 'n' roll." She was still in her 20s when the show was made. In an interview she described her experience:

> More than once, I'd be trying to follow up with a kid (featured in the show) and I would discover pretty quickly that the *only* way I could elicit a response was through a text message or social networking site. I would place call after call, or send e-mail after e-mail—nothing. But with a text, or a message on Facebook, a response would ping back within minutes.
>
> This phenomenon was a surprise; it made me feel old-fashioned—and old. I thought my experience would resemble that of the kids more than their parents, as I'm not a parent yet and certainly

still empathize with being someone's child. The majority of teenagers we talked to expressed good-natured exasperation that their parents "didn't know how to work a computer" or barely understood text messaging. I was confident that because I'm completely comfortable using a computer, e-mail, and a cell phone, I'd relate pretty quickly to how the kids we met communicate online. This was not the case.[4]

Our Digital Addiction

Starbucks has for some time marketed its coffee shops as the "third place," a place where people hang out beyond their homes and offices. Part of the allure was wi-fi availability in most of its stores. Apparently, it has become the first place for some of its customers. Starting in August 2011, some busy Starbucks coffee shops in New York City have started blocking electrical outlets to discourage laptop users from hogging space and to free up seats for other customers.[5]

In 2011 in the UK, a 20-year-old Xbox player died from a clot suspected to be deep vein thrombosis typically associated with lack of mobility on long plane flights. His father said he would often play for 12 hours at a time.[6] In 2005, a South Korean player died after a marathon three-day gaming session. On its Xbox Live site, Microsoft has a lengthy "Healthy Gaming Guide," which recommends a healthy lifestyle, taking frequent breaks, and correct body postures, among other advice.[7]

In 2010 Nintendo put a notice on its site in Japan saying that "children under the age of seven should not use 3-D games, since their eyes are not fully developed. It also says gamers of any age should not play the 3Ds in 3-D mode for longer than 30 minutes at a time. For 2-D game play, Nintendo recommends breaks after an hour of play."[8]

Websites like NetAddiction.com offer self-assessment tests to determine whether technology has become a drug. Among the questions used to identify those at risk: Are you frequently checking your email? Do you often lose sleep because you log in late at night? If you answered "often" or "always," it says technology may be taking a toll on you.[9]

reSTART, an Internet addiction recovery center, says its research is showing the perils of the digital age: addiction, distraction, immaturity, disconnection, lower empathy, lowered creativity and analytical abilities, and increased depression.[10] It offers a 45-day "detox" program at a Retreat Center situated in a "serene natural environment in rural Fall City, Washington, in the Pacific Northwest."

Our Digital Fingerprints

In 2009, *Wired* magazine writer Evan Ratliff "vanished," and the magazine offered a bounty of $5,000 to anyone who could find him. The magazine posted clues as to his whereabouts, and it was fascinating to watch:

> What had started as an exercise in escape quickly became a cross between a massively multiplayer online game and a reality show. A staggeringly large community arose spontaneously, splintered into organized groups, and set to work turning over every rock in Ratliff's life.[11]

What was scary about the whole exercise was how many digital fingerprints Ratliff was leaving even as he was trying to stay underground. Even scarier was how even amateur sleuths were ingenious enough to trace him.

In comparison, there are professional sleuths like the start-up Social Intelligence, which generates reports on job applicants or monitors existing employees. It does so based on employer predefined criteria, both positive and negative. "Negative examples include racist remarks or activities, sexually explicit photos or videos, and illegal activity such as drug use. Positive examples include charitable or volunteer efforts, participation in industry blogs, and external recognition."[12] They search social networks, blog entries, videos, photos, comments, and other forms of user-generated content available publicly on the Internet.

The U.S. Equal Employment Opportunity Commission reports that "75 percent of recruiters are required by their companies to do online research of candidates. And 70 percent of recruiters in the United States report that they have rejected candidates because of information online."[13]

In this chapter's guest column, Professor Mary Cronin writes about "Do Not Track" consumer protections. Even if we can mandate corporations from not tracking us, just about anyone else, curious or malicious, can and will.

Ethical Expectations of Technology Companies

In Chapter 17 we saw how sustainability is a growing expectation of technology companies. It is, however, only a subset of a broader expectation society increasingly has of business ethics.

The Ethisphere Institute is a think tank dedicated to the creation, advancement, and sharing of best practices in business ethics, corporate social responsibility, anti-corruption, and sustainability. Every year it lists the "world's most ethical companies." Attorneys, professors, government officials, and organization leaders assisted Ethisphere in creating the scoring methodology.

The 2011 list includes several technology companies including Accenture, Adobe Systems, Avaya, Becton Dickinson, Cisco Systems, Juniper Networks, Microsoft, Philips, Salesforce.com, Singapore Telecom, Swisscom, Symantec Corporation, Teradata Corporation, and T-Mobile USA.[14]

Don Rickert has a broader definition of ethics. In a blog post he writes,

> While the various professional organizations to which people involved in technology might belong have codes of ethics, serious reflection on the ethical impact of technology in the workplace is a rare thing.[15]

To expand on Rickert's point, very few technical or business schools put much emphasis on ethics in their curricula. And when they do, like Professor Herman Tavani who teaches computer ethics at Rivier College in Nashua, NH, and has written several books on the topic, few technology vendors call him. This walking encyclopedia on cyberethics is not being leveraged enough by technology practitioners. Professor James Moor at Dartmouth, another authority on cyberethics, also confirms he is not often consulted by technology vendors.[16]

Needed: A New Generation of Professionals

Given the gap between the technology avalanche and society's ability to absorb it, we will need a new generation of therapists, new career counselors, and new professors to teach ethics to technicians. We will need more people like Kelly Chessen, who is a former suicide hotline counselor and now DriveSavers' official "data crisis counselor."

Part psychiatrist and part tech enthusiast, Chessen's role is to try to calm people down when they lose their digital possessions to failed data drives. Chessen says that some people have gone as far as to threaten suicide over their lost digital possessions and data.

"It's usually indirect threats like, 'I'm not sure what I'm going to do if I can't get the data back,' but sometimes it will be a direct threat such as, 'I may just have to end it if I can't get to the information,'" said Chessen."[17]

We will also need a new generation of human resource professionals who appreciate that younger candidates will have information online that for a generation prior was offline and be careful how they evaluate such social data.

Conclusion

While we can build technologically elite enterprises, we cannot mandate a technologically elite society. The reality is that it is "unelite" and very uneven. So we need a new set of professionals to prepare society for the avalanche of coming technologies. In a guest column, Professor Mary Cronin of Boston College focuses on another challenge we will face in the next few years. She addresses the continuous, automatic, and invisible tracking of individuals by multiple smart devices and related explosion of personal data and the new privacy challenges society will have to address.

Guest Column: Smart Products Consumers Can Trust—Professor Mary Cronin

Mary J. Cronin is a Professor of Information Systems at the Carroll School of Management, Boston College. Dr. Cronin has more than 20 years' experience in managing and advising technology-intensive organizations. During this time she has written extensively about online privacy and data security issues, including the impact of smartphone apps, RFID, smart products, and geotracking.[18]

When I starting writing about online privacy in the 1990s, the declaration that "There is no privacy on the Internet—get over it" was still controversial. Today's upsurge in social networking and the ubiquity of targeted online advertising and customer profiling makes the lack of online privacy abundantly obvious. However, recent developments indicate that the privacy policy pendulum may be swinging back in the direction of online consumer protection.

In December 2010 the FTC issued a report on "Protecting Consumer Privacy in an Era of Rapid Change," which repeated the well-known reality that "many companies—both online and offline—do not adequately address consumer privacy interests." But for the first time the report proposed "a normative framework for how companies should protect consumers' privacy."

In particular the FTC endorsed a browser-based "Do Not Track" mechanism that would provide a simple way for consumers to disable tracking of their online behavior across all Internet sites. Various Do Not Track bills that would give teeth to the FTC framework have been working their way through Congress and state legislatures during 2011. In anticipation of stricter privacy enforcement, Google, Microsoft, and Mozilla are already providing some Do Not Track features in their latest browser releases.

But a browser-based privacy solution won't address the much larger consumer information sharing issues posed by connected devices and products that are called smart products or more generally the Internet of Things. At a time when smart connected products are generating unprecedented amounts of data about the daily lives, locations, health, and habits of consumers in their homes, in their cars, on mobile networks, and in every location that they might visit, the tracking of online browser behavior is just the tip of the iceberg.

Even if Do Not Track becomes the norm for web browsing, the technology and infrastructure of mobile phones and other consumer devices connected to the Internet of Things represent uncharted territory and abundant temptations for product vendors and service providers to test the boundaries of personal privacy. Analysts project that there will be 50 billion smart devices online by 2015—a number that dwarfs Internet-connected computers and the 6 billion mobile subscribers around the globe.

Apps designed for smartphones with GPS chips already track consumer locations 24 hours a day, often reporting this information to third-party partners who mine it for insights about consumer habits and for precision targeting of individuals. In addition to tracking consumer location, smartphone apps can access the contact information in subscriber phone books, the photographs, and other media stored on the device, and other uses of the phone.

As more smart products are connected to the Internet, new types of reporting and analysis of consumer behavior are enabled. And unlike web tracking, most smart product data collection operates behind the scenes in ways that are invisible and largely unfamiliar to the product owner. What's more, smart product connections that enable monitoring are typically embedded in appliances, automobiles, and medical devices that may not provide any two-way communication options for the consumer. Often data collection is built into the functionality of the smart product itself so that the owner cannot effectively use the device without a connection that also enables vendor reporting and oversight. By design, there is currently no way to set a smart product Do Not Track option, even temporarily.

As I noted in my book *Smart Products, Smarter Services*, "the features that make smart products so intelligent are often the same ones that make them capable of amassing significantly more data about more aspects of their owner's life than ever before." From the moment the buyer activates a smart product, that device monitors its environment and often the presence of nearby connected devices on the same network, often noting any unauthorized connections or uses. It records the exact times and dates that software and content are accessed, scans for upgrades or vendor-controlled instructions, and uses this data to allow or restrict certain types of use by the consumer. Such monitoring and reporting may sound like a futuristic scenario but it's as close as your Kindle, your smart thermostat, or your car's cruise control and collision avoidance systems.

The major new privacy challenge for this decade is the continuous, automatic, and invisible tracking of individuals by multiple smart devices. In the aggregate, consumer-owned smart products collect and report data at a level of precision and frequency that vastly outstrips the consumer information collected online.

The number of smart products owned by a typical consumer is growing at a rapid pace. Home health monitoring and smart energy systems for the home are still in an early adoption stage; by the end of this decade they are likely to be as common as Internet-connected TVs and smartphones in middle-class U.S. households. Smart devices will inevitably develop more sophisticated behavioral tracking and data reporting capabilities and, in the absence of privacy guidelines, vendors will use those capabilities to the fullest. Until smart product privacy gets more attention, consumers are unlikely to realize the extent of tracking and highly personal data collection that is enabled by the smart products used in their daily routines.

Smart Products That Consumers Can Trust—A Business Opportunity

Rather than waiting for the government to mandate smart product policies, companies would be well served in adopting and disclosing consistent and verifiable guidelines for the permissible use, duration of storage, and data security protection measures taken to protect all of the consumer information that is collected by their connected products. Smartphone, automotive, home entertainment, and other smart product vendors should also disclose the privacy practices of their ecosystem partners who develop applications and peripherals for their products.

Providing clear information to consumers about smart product data collection capabilities and offering buyers a spectrum of service and privacy options can provide new business opportunities for vendors. Leveraging the communications capabilities of smart products in ways that allow owners to talk directly to the vendors will create a new form of value and encourage even more data sharing. Many consumers would opt to accept the vendor's stated data collection processes, in exchange for value-added services, improved customer support, and the personalization that such data collection enables. Those who opt out might do so only temporarily. Having control and choice about data sharing would make buyers more comfortable with using the smart product in a variety of ways.

Chapter 20

Market Analysts Morphing

At the end of this chapter, we present the Annual Shareholder Letter that Jeffrey P. Bezos, CEO of Amazon, sent to investors. It starts with "Random forests, naïve Bayesian estimators, RESTful services, gossip protocols, eventual consistency, data sharding, anti-entropy, Byzantine quorum, erasure coding, vector clocks . . ."

Huh? This is from a retailer?

Amazon has good reason to talk technology and related math and science. It is the largest online retailer.[1] Its Kindle, eBook sales, and Amazon Web Services revenues make up 10 percent of its revenues and are some of the fastest growing products in its store. Nonetheless, stop and ask: How many Wall Street analysts can relate to the language that Bezos used?

The Geeks on Wall Street

Wall Street spends plenty on technology. Much has been written about the "quants," especially during the market meltdown of 2008. Scott

Patterson of the *Wall Street Journal,* who wrote a book on the topic, says about them:

> By the early 2000s, such tech-savvy investors had come to dominate Wall Street, helped by theoretical breakthroughs in the application of mathematics to financial markets, advances that had earned their discoverers several shelves of Nobel Prizes.[2]

Yet, even as Wall Street is investing massively in its own technologies, its analysis of technology of industries it follows has not been keeping pace.

Early in 2011, Amazon stock took a hammering even as the company announced robust growth—net income grew an impressive 28 percent over the year ended December 31, 2010, but it also reported lower margins. Evan Schuman, who covers the retail sector, wrote, "It's really about Wall Street's difficulty in understanding IT investments."[3] He went on to add, ". . . its actions with Amazon a few days ago really raise serious doubts about whether any Wall Street trader should be allowed to have a driver's license."

Amazon's sin was that its capital expenditures (capex) grew from around $373 million in 2009 to $979 million in 2010. As a percentage of revenues, capex increased from 1.5 percent in 2009 to 2.9 percent in 2010. Schuman's exasperation was related to Wall Street's failure to understand it was due to investment in fulfillment centers—up from 39 to 52, a 33 percent increase, in 2010. Additionally, the company is adding technology infrastructure capacity to support its fast growing cloud computing business and Kindle manufacturing.

Amazon engaged in another session of analyst soul-searching after a significant outage in its cloud services in April of 2011. The *blueGecko* blog talked about widespread ignorance about cloud computing and used as an example an article in the *Wall Street Journal.* Quoting the authors:

> Here the authors seem to be referring to EC2 availability zones. As most who have worked even a little with EC2 know, when you run an instance or store volumes in one availability zone, there is no automatic mechanism available to "reroute capacity" between availability zones.[4]

Financial Analysts and Domain Knowledge

Dr. Paul Kedrosky is an investor, author, entrepreneur, and a technology analyst for CNBC. On his blog, *Infectious Greed*, and in presentations at various industry events, he covers a wide range of technology and financial topics.

His perspective on financial analysts keeping up with technology:

> Wall Street's sell-side (those that work for investment banks and publish much of the stock research we see) has a very low competency hurdle when it comes to domain expertise. As long as your results are nontoxic (that is, you make money for the firm), and your clients and trading desk don't hate you, that's all that matters.
>
> There are some very good analysts with deep domain expertise who make terrible picks, and there are very bad analysts with no/errant domain expertise who make good picks. Further, the financial side of the Street more than trumps the industry side, so being able to geek out about thin-film poly-Si is much less important than having a developed cash flow model from 2011–2015.
>
> On the buy side, things are very different. Long- and short-side analysts from fundamentally driven firms have fantastically smart analysts who generally know a great deal about domains. The reason, of course, is that there is direct feedback to them from their investments. They buy and sell stuff, and they are held accountable for real portfolios, so they have to be able to bridge industry, financial, and picking expertise.
>
> None of this is to say that there aren't sell-side analysts capable of same—because there are many of them. It's just that the incentive structure doesn't encourage that kind of specialization (nor does it actively discourage it), with the result being that it's a coin-flip what sort of expertise you run across on the sell-side of Wall Street when it comes to highly technical domains.

A fund manager goes even further:

> What is a sell-side researcher's job? It's not to know the most about all aspects of the company (even though we would like to think so). It's to know enough about what *moves the stock* and also how to build a reputation for conveying that information to clients who will pay for their perspective.

> And even on the buy-side, knowing too much about a particular company or sector can IMPAIR your investment returns. Sometimes it's easier to really zero in on "what matters" in terms of a company and its stock price when it's greenfield territory, versus something where you have all the nooks and crannies figured out.

So, financial analysts end up balancing "knowing too much" against the growing tech talk they are increasingly presented. Mark Little, Senior Vice President of GE Global Research, uses terms like Biomimetics in presentations to financial analysts. That refers to the discipline of science mimicking nature, as in GE drawing innovative inspiration for moisture repellants from the lotus leaf. As we saw in the UPS case study in Chapter 1, its CEO tells investors "we're about half a transportation company, half a technology company."

As we saw in Chapter 2, Tony Prophet of HP explains to financial analysts the gory details of its global technology supply chain. Then in Chapter 18, we saw David Meline, CFO of 3M, present to Wall Street a slide on 3M's "periodic table" that summarizes 46 different areas in which it offers technology products.

As more companies talk that way to investors, the more tech-savvy Wall Street analysts will have to become. In the end, behind all that technology are financial numbers. As Bezos points out in his letter: "Now, if the eyes of some shareowners dutifully reading this letter are by this point glazing over, I will awaken you by pointing out that, in my opinion, these techniques are not idly pursued—they lead directly to free cash flow."

The Evolving Industry Analyst

Gartner, the technology research firm (different from Wall Street analysts), covered the Amazon outage mentioned earlier extensively and has reports for its clients with titles such as "Protecting Sensitive Data in Amazon EC2 Deployments."

But search the Gartner database for 3M and you get very few hits. In Chapter 18 we saw 3M has products in 46 different technology platforms. Gartner is the largest technology research firm. What gives?

The reality is that the industry analyst marketplace is itself morphing. Over the course of the book we have mentioned firms like Redmonk, Horses for Sources, iSuppli, and other specialized analyst firms. In the Lexmark Genesis story in Chapter 16, we saw how they targeted technology blogs for their product launch, rather than traditional technology media and analysts.

Evangelos Simoudis, a Senior Managing Director at the venture capital firm Trident Capital, says, "We are a Forrester customer and were a Gartner customer. When it comes to emerging areas for our investments like the social Internet, we use firms like Altimeter Group (which has several Forrester alumni), as well as independent analysts like Esteban Kolsky (who is ex-Gartner), who we see as better "plugged in" to the ecosystem and the market's dynamics. We find that when it comes to business executives driving a decision around emerging technologies the larger, traditional analyst firms are still lacking the market's pulse."

Chris Selland, a former analyst at Yankee Group in the 1990s and later at Aberdeen Group (both industry analyst firms), says:

> During the '90s, there was a tremendous thirst among corporate executives for better understanding of how they could leverage technology and particularly this newfangled "World Wide Web" thing.
>
> During the past 10 years that's been changing dramatically. The typical executive today has a much better understanding of technology, and much less motivation or desire to read a 100-page report that takes a team of analysts six months to develop.
>
> Today's best "analysts"—those who deliver the most true insight and perspective—are almost entirely individuals, not big brand-name firms. Many models for confederating analysts and influencers are being brought to market. I participate in two confederations of independents: the Enterprise Irregulars, which is a loosely connected discussion group for many [former big-firm analysts (including the author of the book)], and Focus Research, which has applied a community and Q&A model to the market.

Frank Scavo of Computer Economics, introduced in Chapter 18, says:

> At first glance, the IT analyst business would appear to be consolidating, with Gartner, Forrester, and IDC having acquired many of the

smaller firms over the past decade. Beneath the surface, however, there has been an explosion of independent analysts and small analyst firms with a laser-like focus on certain markets and industries.

I first started blogging as the Enterprise System Spectator in 2002, as an experiment to see whether blogging would be useful in conjunction with my consulting work at Strativa, a management consulting firm I co-founded in 2000. At the time there were only a handful of tech bloggers, mostly consumer-tech-oriented blogs, such as *Engadget*, and only a few like me focused on enterprise IT. In fact, in 2004, I was voted as one of the top 10 independent tech bloggers. That just goes to show you how small the field was back then.

Today, there are hundreds if not thousands of tech bloggers and small analyst firms that provide an independent voice and counterweight to the traditional analyst firms. My early experiment in blogging got me involved as a contributing analyst to Computer Economics and our eventual acquisition of that 30-year-old IT research firm. It also put me in contact with many like-minded independent analysts around the world, with whom I collaborate. In 2010, some of us joined together to form Constellation Research, under the leadership of R. "Ray" Wang, a former Forrester analyst.

So, on the one hand, there may be only a few major analyst firms left standing. On the other hand, there are many new sources of expertise. Some of the best insights are coming from many voices that are enabled by blogging and other social media, though it does take some effort on the part of buyers to find them.

Conclusion

As the technology elite we profile in the book, such as Amazon, 3M, HP, and UPS, pioneer ways of communicating their own technology prowess to financial and industry analysts, they raise the bar for those analysts. They also raise it for peer companies they compete with or are benchmarked against. In the next section we present the complete text of a groundbreaking letter that Bezos sent to his shareholders.

Case Study: Amazon 2010 Shareholder Letter

Below we present the complete text of a letter Jeffrey P. Bezos, CEO of Amazon, sent to his shareholders, and by proxy to financial and industry analysts. It is groundbreaking in that it does not hesitate to use technology and math terms you would typically hear at a university lecture or in a scientific forum. The language raises the bar for financial and industry analysts who track Amazon. It also raises expectations of Amazon peers and competitors to be as sophisticated in use of technology and in their external communications about their technologies.

To our shareowners:

Random forests, naïve Bayesian estimators, RESTful services, gossip protocols, eventual consistency, data sharding, anti-entropy, Byzantine quorum, erasure coding, vector clocks . . . walk into certain Amazon meetings, and you may momentarily think you've stumbled into a computer science lecture.

Look inside a current textbook on software architecture, and you'll find few patterns that we don't apply at Amazon. We use high-performance transactions systems, complex rendering and object caching, workflow and queuing systems, business intelligence and data analytics, machine learning and pattern recognition, neural networks and probabilistic decision making, and a wide variety of other techniques. And while many of our systems are based on the latest in computer science research, this often hasn't been sufficient: our architects and engineers have had to advance research in directions that no academic had yet taken. Many of the problems we face have no textbook solutions, and so we—happily—invent new approaches.

Our technologies are almost exclusively implemented as services: bits of logic that encapsulate the data they operate on and provide hardened interfaces as the only way to access their functionality. This approach reduces side effects and allows services to evolve at their own pace without impacting the other components of the overall system. Service-oriented architecture—or SOA—is the fundamental building abstraction for Amazon technologies. Thanks to a thoughtful and far-sighted team of engineers and architects, this approach was applied at

Amazon long before SOA became a buzzword in the industry. Our e-commerce platform is composed of a federation of hundreds of software services that work in concert to deliver functionality ranging from recommendations to order fulfillment to inventory tracking. For example, to construct a product detail page for a customer visiting Amazon.com, our software calls on between 200 and 300 services to present a highly personalized experience for that customer.

State management is the heart of any system that needs to grow to a very large size. Many years ago, Amazon's requirements reached a point where many of our systems could no longer be served by any commercial solution: our key data services store many petabytes of data and handle millions of requests per second. To meet these demanding and unusual requirements, we've developed several alternative, purpose-built persistence solutions, including our own key-value store and single table store. To do so, we've leaned heavily on the core principles from the distributed systems and database research communities and invented from there. The storage systems we've pioneered demonstrate extreme scalability while maintaining tight control over performance, availability, and cost. To achieve their ultra-scale properties these systems take a novel approach to data update management: by relaxing the synchronization requirements of updates that need to be disseminated to large numbers of replicas, these systems are able to survive under the harshest performance and availability conditions. These implementations are based on the concept of eventual consistency. The advances in data management developed by Amazon engineers have been the starting point for the architectures underneath the cloud storage and data management services offered by Amazon Web Services (AWS). For example, our Simple Storage Service, Elastic Block Store, and SimpleDB all derive their basic architecture from unique Amazon technologies.

Other areas of Amazon's business face similarly complex data processing and decision problems, such as product data ingestion and categorization, demand forecasting, inventory allocation, and fraud detection. Rule-based systems can be used successfully, but they can be hard to maintain and can become brittle over time. In many cases, advanced machine learning techniques provide more accurate classification and can self-heal to adapt to changing conditions. For example, our search engine employs data mining and machine learning algorithms that run in the background to build topic models, and we apply information extraction algorithms to identify attributes and extract entities from unstructured descriptions, allowing customers to narrow their searches and quickly find the desired product. We consider a large number of factors in search relevance to predict the probability of a customer's interest and optimize the ranking of results. The diversity of products demands that we employ modern regression techniques like trained random forests of decision trees to flexibly incorporate thousands of product attributes at rank time. The end result of all this behind-the-scenes software? Fast, accurate search results that help you find what you want.

All the effort we put into technology might not matter that much if we kept technology off to the side in some sort of R&D department, but we don't take that approach. Technology infuses all of our teams, all of our processes, our decision-making, and our approach to innovation in each of our businesses. It is deeply integrated into everything we do.

One example is Whispersync, our Kindle service designed to ensure that everywhere you go, no matter what devices you have with you, you can access your reading library and all of your highlights, notes, and bookmarks, all in sync across your Kindle devices and mobile apps. The technical challenge is making this a reality for millions of Kindle owners, with hundreds of millions of books, and hundreds of device types, living in over 100 countries around the world—at 24 × 7 reliability. At the heart of Whispersync is an eventually consistent replicated data store, with application defined conflict resolution that must and can deal with device isolation lasting weeks or longer. As a Kindle customer, of course, we hide all this technology from you. So when you open your Kindle, it's in sync and on the right page. To paraphrase Arthur C. Clarke, like any sufficiently advanced technology, it's indistinguishable from magic.

Now, if the eyes of some shareowners dutifully reading this letter are by this point glazing over, I will awaken you by pointing out that, in my opinion, these techniques are not idly pursued—they lead directly to free cash flow.

We live in an era of extraordinary increases in available bandwidth, disk space, and processing power, all of which continue to get cheap fast. We have on our team some of the most sophisticated technologists in the world—helping to solve challenges that are right on the edge of what's possible today. As I've discussed many times before, we have unshakable conviction that the long-term interests of shareowners are perfectly aligned with the interests of customers.

And we like it that way. Invention is in our DNA and technology is the fundamental tool we wield to evolve and improve every aspect of the experience we provide our customers. We still have a lot to learn, and I expect and hope we'll continue to have so much fun learning it. I take great pride in being part of this team.

As always, I attach a copy of our original 1997 letter. Our approach remains the same, and it's still Day 1.

Jeffrey P. Bezos
Founder and Chief Executive Officer
Amazon.com, Inc

Source: Amazon 2010 Annual Shareholder Letter.

Ready for a quiz based on comprehension of the Bezos letter? Not to worry. If you glossed over the letter or, as Bezos points out, if your eyes are glazed over after reading it, what you need to take away is the techniques and technologies described in the letter are not idly pursued. They lead directly to free cash flow. They are also likely to lead other CEOs to unabashedly talk about their own technologies to their stakeholders.

Endgame: "Welcome to the NFL"

Charlie Feld, whom we introduced in Chapter 15, should be long retired. In the past three decades, as CIO at Frito-Lay, then at the firm he founded, The Feld Group, he influenced IT at countless companies. He has since formed the Feld Group Institute, which tells his view of the technology world on its website:

> The journey of (technology) innovation and sophistication has been nothing short of breathtaking.
>
> What has not been so spectacular has been the evolution and maturing of a management framework that would enable the efficient application of these rich technologies within large enterprises. Every decade these enterprises have become more dependent on IT but the potential still far exceeds the ability of most organizations, industries and governments to harvest IT.[1]

He says, in an interview, IT in most companies has become one-dimensional, and is often just cost or compliance focused. "In the big leagues, you cannot choose to play just defense or just offense. You have to build teams with multiple roles and skills."

"Welcome to the NFL," says Feld.

Elite Technology Teams

The National Football League is a good analogy for building elite technology teams. In the NFL, for any given play a team cannot have more than 11 players on the field. A team roster is 53 players, although on game day only 45 can be declared active. There are offensive units, defensive units, and special teams. There are at least 20 different roles including quarterbacks, guards, defensive ends, and placekickers. There are countless coaches who help hone the skills of the already elite athletes who play at the professional level.

Beyond the IT skills where Feld calls for attention, as we have seen in earlier chapters, an elite technology team needs great industrial designers like Jonathan Ive at Apple. It needs superb talent like Tony Prophet at HP to manage logistics in the complex global technology supply chain. It needs more of those operational geniuses like Jim Miller at Google to manage hyperefficient data centers and real estate in a world where physical presence continues to be hugely important. It needs economists like Todd Stockard at Valence Health who are comfortable with morphing business models and financial executives like Gordon Coburn at Cognizant, who can drive out massive waste in technology spending. It needs attorneys like Benjamin Kern because legal design often trumps product design in technology markets. It needs marketing folks like Jerry Grasso at Lexmark who are socially savvy. It needs "crossover" executives who have experiences at technology vendors and users like Tony Scott at Microsoft and Vijay Ravindran at The Washington Post Co.

The NFL is an appropriate analogy in other ways. The league defines rules for drafting players and recruiting free agents. It has other rules for player decorum on the field and off. It has rules for everything. We are about to enter a phase of more tech-savvy regulators and market watchers. We saw glimpses of government efficiency and innovation in the examples of the country of Estonia, the Hillsborough County Tax

Collector's Office, and Roosevelt Island. As Robert Hoffman points out in Chapter 18, "Tech-savvy policymakers are coming of age. The twentysomethings that first connected Capitol Hill to the information superhighway are now in positions of authority."

NFL commentary has evolved significantly with instant replays, Skycam angles, and other technology. When 3M showcases its "Periodic Table," with its 46 technology platforms, technology analysts can only become savvier. There are plenty of blogs and smaller specialty analyst firms that are defining new rules for the watched and the market watchers.

The NFL analogy rings true even more after the 2011 owner-player dispute. As the months-long lockout ended, the Players Association issued a video featuring 13 players from various teams delivering personal messages to fans thanking them for their patience and support.[2]

The society surrounding the NFL with fans and their "fantasy leagues," the players union, the cities that host the stadiums, and the charities the teams support all significantly influence the game.

In the same way, the society around technology is complex and varied. There are thorny ethical, legal, and other issues we will increasingly have to factor in to our technology design.

Any Given Sunday

Chapter 20 showed a letter that Jeff Bezos, CEO of Amazon, sent to his shareholders and it is full of technology terms and trends. Amazon is definitely a technology elite company. It is the largest online retailer.[3] Its Kindle, eBook sales, and Amazon Web Services revenues make up 10 percent of its revenues and are some of the fastest growing products in its store. It would be fair to say that Amazon focuses today as much on Apple and Google, as it does on Walmart or Sears or Barnes & Noble. To net it out, Amazon has a wide range of attributes that allow it to compete on multiple fronts: offense, defense, and special teams.

The seesaw battle between Amazon and Walmart has been fascinating to watch. In 1999 Walmart sued then fledgling Amazon for hiring several of its technology and logistics executives. Nonetheless, Walmart seemed unstoppable. *BusinessWeek* asked in 2002, "Will Wal-Mart Take

Over the World?," giving credit to its extremely sophisticated use of information technology.[4]

By 2007, that technology advantage had shrunk. A technology fund manager observed, "For years, Walmart was held up as a shining example of cutting-edge thinking in retail technology. But today, when I hear about a retailer doing something cutting edge, it's never Walmart being talked about."[5]

For a while, Walmart's online business sputtered while Amazon's exploded:

> Walmart's lackluster online history has deep cultural roots. The organization has long been dominated by store managers who feared e-commerce could cannibalize in-store sales, and thus their bonuses, according to a former Walmart.com senior executive.[6]

Now, in a turnaround, Venky Harinarayan and Anand Rajaraman, are running @Walmartlabs "to speed with innovations such as smartphone payment technology, mobile shopping applications, and Twitter-influenced product selection for stores."[7] Yes, these are the same executives who founded Junglee, which Amazon acquired in 1998, and were influential in those formative days at Amazon.

Walmart, Sears, Barnes & Noble, and other traditional retailers are also trying to fight Amazon (and other Internet retailers) by lobbying for taxes on Internet sales (an attractive proposition to many states in these fiscally strapped times) and on sales by its affiliates.[8] They are also wooing its affiliates to join their ecosystems.[9]

And that brings us to another concept the NFL has encouraged—"parity." It has salary caps across teams. Weaker teams get to draft better players the next year. There are few "dynasties."

So, Amazon's advantage may be as fleeting as was Walmart's. Technology may have its own version of parity. This is giving rise to "Phoenix" strategies." In Chapter 3, we saw several examples like Aircell, Delta Airlines, GM, and others where technology is giving them a fighting chance to regain past glory.

The Coming Upheaval

We live in exciting times. To some degree the current landscape is the throwback to the 1960s and 1970s, when we dreamed of competitive advantage through technology. Sabre, the American reservation system, and American Hospital Supply were spoken of fondly for changing their industries. We have a similar opportunity now, but in most organizations the IT group is much more focused on the back office, not on product or revenue or growth.

In reverse, we have technology vendors with an entitlement mindset. Even as Apple and Amazon have shown dramatic business model innovation, too many technology vendors continue with older business models, and feel entitled to their compensation levels. So they cling to 90 percent software gross margins, $5,000 a gallon for printer ink, and 50 percent margins on so-called cheap offshore talent. Either they learn to disrupt themselves or they will get disrupted.

Dramatic changes are needed if you aim to become one of the technology elite.

And a Final Word from Coach Bear Bryant

Coach Paul "Bear" Bryant, one of the most revered football coaches, was quoted as saying: "Show class, have pride, and display character. If you do, winning takes care of itself."

Unfortunately, technology, with the riches and recognition it brings, also seems to bring plenty of arrogance. Coach Bryant's words are more good advice for building an elite technology team.

Conclusion

Beyond the 12 attributes we discussed in Part II, more demanding regulators and market watchers and changing societal expectations are all going to influence our definition of technology elite in the near future. 3M's "Periodic Table," Amazon's technology-rich shareholder letter, UPS's self-definition of being "half a transportation company, half a technology company," Apple's benchmarks for retail excellence are the language the technology elite will increasingly use with many of their stakeholders—their customers, investors, regulators, and society at large.

Notes

Preface

1. www.slideshare.net/vmirchan/ignite-vinnie-mirchandani-sep-2010.
2. *New Florence. New Renaissance* (blog), www.florence20.typepad.com.

Chapter 1

1. 2011 CES Attendee Audit Summary Results, Letter from CEA, www.cesweb.org/docs/2011AuditSummary.pdf.
2. Ed Oswald, "CES 2011: Seven Tablets You Should Know about," betanews, January 7, 2011, www.betanews.com/article/CES-2011-Seven-tablets-you-should-know-about/1294429871.
3. Bill Ford, "Bill Ford Looks Ahead," *Fortune*, May 23, 2011, http://money.cnn.com/magazines/fortune/fortune_archive/2011/05/23/toc.html.
4. Andrew Nusca, "CES: GM Announces OnStar for Rival Brands; Makes In-Vehicle Platform Play," *ZDNet*, January 4, 2011, www.zdnet.com/blog/gadgetreviews/ces-gm-announces-onstar-for-rival-brands-makes-in-vehicle-platform-play/21147.

5. Vinnie Mirchandani, *The New Polymath* (Hoboken, NJ: John Wiley & Sons, 2010).
6. Matt Murphy and Mary Meeker, "Top Mobile Internet Trends," February 10, 2011, www.slideshare.net/kleinerperkins/kpcb-top-10-mobile-trends-feb-2011.
7. Matthew Shaer and Mark Zuckerberg, "Kids under 13 Should Be Welcome on Facebook, Too," *The Christian Science Monitor,* May 25, 2011, www.csmonitor.com/Innovation/Horizons/2011/0525/Mark-Zuckerberg-Kids-under-13-should-be-welcome-on-Facebook-too.
8. "Nintendo Policy," www.nintendo.com/corp/privacy.jsp.
9. Walter Hamilton, "Elder Care Goes High Tech," *Los Angeles Times,* June 17, 2011, www.latimes.com/business/la-fi-boomer-homes-20110617,0,7872478,full.story.
10. Matthew Danzico, "Cult of Less: Living Out of a Hard Drive," *BBC News*, August 16, 2010, www.bbc.co.uk/news/world-us-canada-10928032.
11. "Creating a Bring Your Own Technology (BYOT) Program," IT Business Edge, www.itbusinessedge.com/slideshows/show.aspx?c=84802&utm_source=itbe&utm_medium=email&utm_campaign=ISR&nr=ISR.
12. Kyle Stack, "In-Chest Sensors Gather Data on NFL Prospects," *Wired,* February 23, 2011, www.wired.com/playbook/2011/02/nfl-combine-chest-sensors.
13. Do restaurant website, www.doattheview.com.
14. Paul Eng, "State Farm Insurance Offers Drivers an OnStar-Like System," *Consumer Reports,* August 5, 2011, http://news.consumerreports.org/cars/2011/08/state-farm-insurance-offers-drivers-an-onstar-like-system.html.
15. Christine Kenneally, "How to Fix 911," *Time,* April 17, 2011, www.time.com/time/magazine/article/0,9171,2062452,00.html#ixzz1ITq5J6Dd.
16. Jennifer Van Grove, "How the Humble Washing Machine Is Getting a Digital Spin," *Mashable*, August 9, 2011, http://mashable.com/2011/08/09/digital-washing-machine.
17. "Technology Facts," UPS Pressroom, http://pressroom.ups.com/Fact+Sheets/Technology+Facts%3A+UPS.
18. "Partners in Leadership: Delivering Technology and Innovation," UPS Pressroom, www.pressroom.ups.com/About+UPS/UPS+Leadership/Speeches/David+Barnes/ci.Partners+in+Leadership%3A+Delivering+Technology+and+Innovation.print.
19. Dave Demerjian, "Clean(er), Quiet(er) Landings Coming to an Airport Near You," *Wired,* May 21, 2008, available at www.wired.com/autopia/2008/05/flying-circles.

Chapter 2

1. Kevin O'Marah and Debra Hoffman, "The AMR Supply Chain Top 25 for 2010," Gartner Research, June 2, 2010, www.gartner.com/Display Document?id=1379613.
2. Paul McDougall, "iPad Is Top Selling Tech Gadget Ever," *InformationWeek*, October 7, 2010, www.informationweek.com/news/storage/portable/227700347.
3. Dan Gilmore, "The New Supply Chain Lessons from Dell," *Supply Chain Digest*, April 10, 2008, www.scdigest.com/assets/FirstThoughts/08-04-10.php.
4. Adam Lashinsky, "Inside Apple," *Fortune*, May 23, 2011, http://money.cnn.com/magazines/fortune/fortune_archive/2011/05/23/toc.html.
5. Charlie Babcock, "Lessons from Farmville," *InformationWeek*, May 26, 2011, http://analytics.informationweek.com/abstract/10/7254/Social%20Networking-Collaboration/lessons-from-farmville.html?cid=nl_rep__iwkrnwls l0531 2011.
6. Amazon 2010 Shareholder letter available at company website, http://phx.corporate-ir.net/External.File?item=UGFyZW50SUQ9OTA4ODB8Q2hpbGRJRD0tMXxUeXBlPTM=&t=1.
7. "Kindle to Generate $5.42 bln Revenue in 2011 for Amazon: Analyst," *International Business Times*, May 10, 2011, www.ibtimes.com/articles/143318/20110510/amazon-com-kindle-ebook-reader-device-apps-video-book-titles-store-nasdaq-stock-market-wi-fi-114.htm.
8. Nick Wingfield, "Amazon Prospers on the Web by Following Wal-Mart's Lead," *Wall Street Journal*, November 22, 2002, http://courses.washington.edu/inde101/Class%20Notes/Amazon%20Follows%20Wal-Mart%27s%20Lead.htm.
9. Katie Fehrenbacher, "Google: There's No Magic Needed for Greener Data Centers, *Earth2Tech*, May 25, 2011, www.reuters.com/article/2011/05/25/idUS307015909420110525.
10. "Paypal Q4 2010 Fast Facts," www.paypal-media.com/assets/pdf/fact_sheet/PP_FastFacts_Q410_0111.pdf.
11. Stacey Higginbotham, "Facebook Open Sources Its Servers and Data Centers," *GigaOM*, April 7, 2011, http://gigaom.com/cloud/facebook-open-sources-its-servers-and-data-centers.
12. Brian Solis, "This Just In . . . News No Longer Breaks, It Tweets," *BrianSolis.com* (blog), May 2, 2011, www.briansolis.com/2011/ 05/this-just-in-news-no-longer-breaks-it-tweets.
13. Wade Roush, "Is Twitter Here to Stay?" *MIT Technology Review*, April 6, 2007, www.technologyreview.com/communications/18494/page1.

14. "Twitter's New Search Architecture," Twitter Engineering page, October 6, 2010, http://engineering.twitter.com/ 2010/10/twitters-new-search-architecture.html.
15. Michael Santo, "Amazon.com Beats Google, Apple into Web-Based Music: Cloud Player, Cloud Drive," Examiner.com, March 29, 2011, www.examiner.com/technology-in-national/amazon-com-beats-google-apple-into-web-based-music-cloud-player-cloud-drive#ixzz1NOqpud9z.
16. Larry Dignan, "HP's PC Supply Chain by the Numbers," *ZDNet*, June 10, 2010, www.zdnet.com/blog/btl/hps-pc-supply-chain-by-the-numbers/35698.
17. Steve Lohr, "Stress Test for the Global Supply Chain," *New York Times,* March 19, 2011, www.nytimes.com/2011/03/20/business/20supply.html.
18. Transcript—Hewlett-Packard at Raymond James IT Supply Chain Conference, Thomson StreetEvents, http://phx.corporate-ir.net/External.File?item=UGFyZW50SUQ9MzQ5ODI1OXxDaGlsZElEPTQwNzg0MnxUeXBlPTI=&t=1.
19. Wei Gu, "China's Growth Hinges on Wild West," *Reuters*, May 31, 2011, http://in.reuters.com/article/2011/05/31/idINIndia-57395120110531.
20. Thomas L. Sims and Jonathan James Schiff, "The Great Western Development Strategy," *China Business Review*, November–December 2000, www.chinabusinessreview.com/public/0011/sims.html.
21. Yantian International Container Terminals, "YICT Holds a Promotion Seminar for Chongqing-Shenzhen Intermodal Service," news release, June 21, 2011, http://61.144.223.202/2006en/newscentre/News.asp?Article_ID=12569.
22. Shipping Online.CN, "Hewlett Packard Exec Checks Out Chongqing Airport Facilities," news release, www.shippingonline.cn/news/newsContent.asp?id=15382
23. Electronic Industry Citizenship Coalition website, www.eicc.info.
24. From HP Global Citizenship website, www.hp.com/hpinfo/globalcitizenship/environment/materials.html.
25. From HP Global Citizenship website, www.hp.com/hpinfo/globalcitizenship/society/conflict_minerals.html.

Chapter 3

1. Jonathan Cohn and Mark Robson, "Taming Information Technology Risk: A New Framework for Boards of Directors," Oliver Wyman, www.oliverwyman.com/ow/pdf_files/OW_EN_GRC_2011_PUBL_Taming_IT_Risk.pdf.

2. James Surowiecki, "The Next Level," *The New Yorker,* October 18, 2010, www.newyorker.com/talk/financial/2010/10/18/101018ta_talk_surowiecki #ixzz1NhgZwOQh.

3. "The State of Smarter Industries: Barcelona 2010 Industry Reports. IBM, www-935.ibm.com/services/us/gbs/industries/symposiumreports.

4. "How Craigslist, Yahoo, and Monster Killed Newspapers," *In-Depth Technology Market Research*, March 15, 2011, www.in-depthresearch.com/ marketing-trends/how-craigslist-yahoo-and-monster-killed-newspapers.

5. David Stevenson, "Dig That Moat Around Your Portfolio," *Investor's Chronicle UK*, September 1, 2008, http://goliath.ecnext.com/coms2/gi_0199 -8544269/Dig-that-moat-around-your.html.

6. Chris Ward, "Warren Buffett Hesitant on Apple, Tech Companies, *The Unofficial Apple Weblog*, March 22, 2011, www.tuaw.com/2011/03/22/ warren-buffet-hesitant-on-apple-tech-companies.

7. Mary Weier Hayes, "Coke's RFID-Based Dispensers Redefine Business Intelligence," *Information Week,* June 6, 2009, www.informationweek.com/ news/mobility/RFID/217701971.

8. Emily Thornton, "Burlington Northern and the Revival of Railroads," *Bloomberg Businessweek,* October 23, 2008, www.businessweek.com/magazine/ content/08_44/b4106058122336_page_2.htm.

9. Lennie Bennett, "Leaks at the Dali Museum in St. Petersburg Called Tiny Flaws," Tampabay.com, May 11, 2011, www.tampabay.com/ news/leaks-at -the-dali-museum-in-st-petersburg-a-tiny-flaw/1168800.

10. Michelle Manchir, "Library of the Future: Wi-fi, Flat Screens, Automated Book Sorting," *Chicago Tribune*, March 7, 2011, http://articles .chicagotribune.com/2011-03-07/news/ct-met-library-of-the-future-0308 -20110307_1_wi- fi-new-library-library-district.

11. "Newport Beach Loves Books," Newport P.J., Orange County's Trusted Professional Investigators, March 30, 2011, www.newportpi.com/newport -beach-loves-books.

12. Steven Pearlstein, "Mark Them Tardy to the Revolution," *The Washington Post,* May 24, 2011, www.washingtonpost.com/steven-pearlstein-mark-them -tardy-to-the-revolution/2011/05/24/AG1vKYDH_story.html.

13. Kosaka Smelting and Refining, "For Turning Old Cell Phones into Gold Mines," *Fast Company*, March 2011, www.fastcompany.com/ most-innovative-companies/2011/profile/kosaka-smelting-and-refining.php.

14. Jacob Aron, "3D Printing: Game Add-On Makes Minecraft Edifices Real," *New Scientist*, August 4, 2011, www.newscientist.com/article/mg21128236 .200-3d-printing-game-addon-makes-minecraft-edifices-real.html.

15. Karen Klein, "Trademarkia Upends Trademark Filing Process, Angers IP Lawyers," Bloomberg, May 31, 2011, www.bloomberg.com/news/2011-05-31/trademarkia-upends-trademark-filing-process-angers-ip-lawyers.html?campaign_id=mag_Jun2&link_position=link31.
16. Claire Miller Cain, "Computers Studies Made Cool on Film and Now on Campus," *New York Times*, June 10, 2011, www.nytimes.com/2011/06/11/technology/11computing.html?nl=todaysheadlines&emc=tha25.
17. John Markoff, "Michael Dell Should Eat His Words, Apple Chief Suggests," *New York Times*, January 16, 2011, www.nytimes.com/2006/01/16/technology/16apple.html.
18. Chariesse Jones, "Delta Brings Back Red Coats to Help with Customer Service," *USA Today,* June 25, 2011, www.usatoday.com/money/industries/travel/2009-06-24-delta-red-coats-service_N.htm.
19. Devin Leonard, "The U.S. Postal Service Nears Collapse," *Bloomberg Businessweek,* May 26, 2011, www.businessweek.com/magazine/content/11_23/b4231060885070.htm.
20. Sharon Bertsch McGrayne, *The Theory That Would Not Die* (New Haven, CT: Yale University Press, 2011).
21. "The One Prize 2011 Open International Design Competition to Envision the Sixth Borough of New York City," www.oneprize.org/1about.html.
22. Verdant Power, "The RITE Project: East River—New York, NY," May 12, 2011, http://verdantpower.com/wp-content/themes/Verdant/downloads/VerdantPower_RITE.pdf.
23. "Fast Trash!" *Urban Omnibus*, May 12, 2010, http://urbanomnibus.net/2010/05/fast-trash.
24. Richard Perez-Pena, "For Cornell Tech School, a $350 Million Gift From a Single Donor", *New York Times*, December 19, 2011 http://www.nytimes.com/2011/12/20/nyregion/cornell-and-technion-israel-chosen-to-build-science-school-in-new-york-city.html?_r=1
25. Michael Grynbaum, "East River Ferry Service, with 7 Stops, Starts Run," *New York Times,* June 13, 2011, www.nytimes.com/2011/06/14/nyregion/east-river-ferry-service-begins-with-7-stops.html.
26. Nick Chambers, "Leviton's Portable Electric Car Charging Stations Are a Big Deal, Here's Why," Plugincars, March 7, 2011, www.plugincars.com/levitons-portable-electric-car-charging-station-big-deal-heres-why-106905.html.
27. Franchise Solutions, "We Empower Opportunity," http://newdeal.feri.org/speeches/1932d.htm.

Chapter 4

1. Yuqing Xing and Neal Detert, "How the iPhone Widens the United States Trade Deficit with the People's Republic of China," ADBI Working Paper Series, ADB Institute, May 2011, www.adbi.org/files/2010.12.14.wp257.iphone.widens.us.trade.deficit.prc.pdf.
2. "Global Rollout of 3G-Ready iPhone Kicks Off," *The Economic Times,* July 11, 2008, http://m.economictimes.com/PDAET/articleshow/3222750.cms.
3. Shira Ovide, "Groupon IPO: It's Here!" *Deal Journal*, June 2, 2011, http://blogs.wsj.com/deals/2011/06/02/groupon-ipo-its-here.
4. Will Richmond, "Netflix Expands to 43 Latin American Countries But Faces New Broadband Challenges," *Video Nuze,* July 5, 2011, www.videonuze.com/blogs/?2011-07-05/Netflix-Expands-to-43-Latin-American-Countries-But-Faces-New-Broadband-Challenges/&id=3126.
5. The World Is Flat: A Brief History of the Twenty-First Century," Bookshelf, *New York Times*, www.thomaslfriedman.com/bookshelf/the-world-is-flat.
6. Pankaj Ghemawat, "Globalization in the World We Live in Now: World 3.0," *Harvard Business Review*, May 31, 2011, http://blogs.hbr.org/cs/2011/05/globalization_in_the_world_we.html.
7. Sara Corbett, "Can the Cellphone Help End Global Poverty?" *New York Times Magazine,* April 13, 2008, www.nytimes.com/2008/04/13/magazine/13anthropology-t.html.
8. Reena Jana, "Innovation Trickles in a New Direction," *Bloomberg Businessweek,* March 11, 2009, www.businessweek.com/magazine/content/09_12/b4124038287365.htm?chan=innovation_innovation+%2B+design_top+stories.
9. "Frugal Healing," *The Economist*, January 20, 2011, www.economist.com/node/17963427.
10. Tom Edwards, "How to Live Off of Japanese Vending Machines," *Maximum Tech*, May 24, 2011, www.maximumtech.com/how-live-japanese-vending-machines.
11. "No One Uses Voicemail in India and the Concept of Missed Calls," *Rashmi's Blog*, January 9, 2006, http://rashmisinha.com/2006/01/09/no-one-uses-voicemail-in-india-and-the-concept-of-missed-calls.
12. Sultan Qassemi Al Sooud, "Gulf Governments Take to Social Media," *Gulfnews,* May 30, 2011, http://gulfnews.com/opinions/columnists/gulf-governments-take-to-social-media-1.814508.
13. Hoptroff web page, http://hoptroff.com/learn_more.html.

14. Christopher Dann, Sartaz Ahmed, and Owen Ward, "Renewable Energy at a Crossroads," *Strategy+Business,* Autumn 2011, www.strategy-business.com/article/00082?pg=1.

15. Emil Protalinski, "Iceland Taps Facebook to Rewrite Its Constitution," *ZDNet*, June 14, 2011, www.zdnet.com/blog/facebook/iceland-taps-facebook-to-rewrite-its-constitution/1600.

16. SMART, "About SMART," 2010, http://smart.mit.edu/about-smart/about-smart.html?homeintro=2.

17. Sean Dorgan, "How Ireland Became the Celtic Tiger," The Heritage Foundation Leadership for America, June 23, 2006, www.heritage.org/research/reports/2006/06/how-ireland-became-the-celtic-tiger.

18. "South Korea's Pupils to Go Paperless by 2015," *NewScientist Tech*, July 8, 2011, www.newscientist.com/article/mg21128203.400-south-koreas-pupils-to-go-paperless-by-2015.html.

19. "Quincy Cashes in On the Out Cloud," *Bloomberg Businessweek*, May 2, 2011, www.businessweek.com/technology/content/may2011/tc2011052_966792.htm?campaign_id=techn_May3&link_position=link32.

20. Isabelle Pommereau, "Skype's Journey from Tiny Estonian Start-Up to $8.5 Billion Microsoft Buy," *Christian Science Monitor,* May 11, 2011, www.csmonitor.com/World/Europe/2011/0511/Skype-s-journey-from-tiny-Estonian-start-up-to-8.5-billion-Microsoft-buy.

21. Erich Follath, "Tallinn: Estonia's Wired Capital," *Bloomberg Businessweek*, September 5, 2007, www.businessweek.com/globalbiz/content/sep2007/gb2007095_520696.htm.

22. "Estonia's 'Tiger Leap'—Brief Article," *Business Article*, 2011, http://findarticles.com/p/articles/mi_m1309/is_2_37/ai_66579839.

23. "President Ilves: Tiger Leap Should Continue as Citizens' Initiative," Office of the President, February 22, 2010, www.president.ee/en/media/press-releases/2081-president-ilves-tiger-leap-should-continue-as-citizens-initiative/index.html.

24. "e-Estonia: Economy and IT," http://estonia.eu/about-estonia/economy-a-it/e-estonia.html.

25. "A Cyber-Riot," *The Economist,* May 10, 2007, www.economist.com/node/9163598?Story_ID=E1_JTGPVJR.

26. Joshua Davis, "Hackers Take Down the Most Wired Country in Europe," *Wired Magazine*, August 21, 2011, www.wired.com/politics/security/magazine/15-09/ff_estonia?currentPage=all.

27. "Leading the Way," *Government Executive*, August 15, 2011, www.govexec.com/features/0811-15/0811-15s6.htm.

Chapter 5

1. http://solutions.3m.com/wps/portal/3M/en_WW/3M-Tech3/Technology.
2. Vinnie Mirchandani, *The New Polymath* (Hoboken, NJ: John Wiley & Sons, 2010).
3. Thomas Wailgum, "SAP's Not So Secret Weapon: Its Own CIO," *CIO*, December 15, 2011, http://cio.co.nz/cio.nsf/focus/936BEBA0DCFA2860CC2577F90067BFD1.
4. Douglas MacMillan, "Thiel Awards 24 Under-20 Fellowships," *Bloomberg Businessweek*, May 25, 2011, www.businessweek.com/technology/content/may2011/tc20110524_317819.htm.
5. Bernice Yeung, "Silicon Valley Fellowship: Get Ahead by Dropping Out," *Firstpost World*, May 27, 2011, www.firstpost.com/world/thiel-fellowship-16334.html.
6. Christina Warren, "iPod Nano Watch Project Makes Kickstarter History," *Mashable Tech,* December 17, 2010, http://mashable.com/2010/12/17/kickstarter-ipod-nano.
7. Tim Kurkjian, "It's Really Not Great to Be a Switch-Hitter," ESPN MLB, September 3, 2011, http://sports.espn.go.com/mlb/columns/story?columnist=kurkjian_tim&id=5524029.
8. Ibid.
9. http://hst250.history.msu.edu/wiki-online/index.php/Xerox_PARC.
10. Chris Murphy, "IT Must Create Products, Not Just Cut Costs," *Information Week,* Global Week, March 12, 2011, www.informationweek.com/news/global-cio/interviews/229300065?queryText=murphy.
11. "Test Complete: 2011 Ford Edge," *Consumer News,* January 14, 2011, http://news.consumerreports.org/cars/2011/01/test-complete-2011-ford-edge.html.
12. Paul Eisenstein, "Initial Quality of New 2011 Models 'Declined Considerably,' Cautions New Study," *Yahoo Autos*, http://autos.yahoo.com/news/initial-quality-of-new-2011-models-declined-considerably–cautions-new-study.html.

Chapter 6

1. Brad Stone, "Raves for Robert Brunner's All-New Nook," *Bloomberg Businessweek*, June 30, 2011, www.businessweek.com/magazine/raves-for-robert-brunners-allnew-nook-07012011.html.

2. "De Dietrich DTiM1000C 90CM Induction Hob—The Piano, Ultimate Cooking Freedom," *Kitchen Designer Blog*, April 11, 2011, http://thekitchendirectory.wordpress.com/2011/04/14/de-dietrich-dtim1000c-90cm-induction-hob-the-piano-ultimate-cooking-freedom/.
3. Matthew Danzico, "Cult of Less: Living Out of a Hard Drive," *BBC News*, August 16, 2010, www.bbc.co.uk/news/world-us-canada-10928032.
4. Mickey McManus, "Information Liquidity," *Huffington Post,* April 28, 2011, www.huffingtonpost.com/mickey-mcmanus/post_1985_b_854734.html.
5. TedxCMU video, www.youtube.com/watch?v=hFgltc6xi00.
6. Marissa Mayer, "What Comes Next in the Series 13,31,53, 61, 37, 28 . . . ," *The Official Google Blog*, July 3, 2008, http://googleblog.blogspot.com/2008/07/what-comes-next-in-this-series-13-33-53.html.
7. "Google Doodle History," www.google.com/doodle4google/history.html.
8. Dierdre Van Dyk, "The Story of Google Doodles," *Time* Photos, www.time.com/time/photogallery/0,29307,1975305,00.html#ixzz1Px8kubLt.
9. Ibid.
10. Rolleiv Solholm, "The New Oslo Opera House Wins EU Award," *Norway Post,* April 30, 2009, www.norwaypost.no/news/the-new-oslo-opera-house-wins-eu-award.html.
11. Matt Chaban, "Snohetta Takes Broadway with Times Square Repairs," *A/N blog*, http://blog.archpaper.com/wordpress/archives/8200.
12. Snohetta web page, www.snoarc.no/#/about/information/45.
13. "For Design That's Both Social and Beautiful," *Fast Company,* March 2011, www.fastcompany.com/most-innovative-companies/2011/profile/snohetta.php.
14. Johnny Ryan, "Manufacturing 2.0: The Rise of 3-D Printers," *Fortune*, May 23, 2011, http://tech.fortune.cnn.com/2011/05/23/manufacturing-2-0-the-rise-of-3-d-printers.
15. Vinnie Mirchandani, *The New Polymath* (Hoboken, NJ: John Wiley & Sons, 2010).
16. Damon Lavrinc, "Five Mind-Blowing Innovations from Mercedes-Benz," AOL Autos, April 14, 2011, http://autos.aol.com/article/five-tech-innovations-mercedes-benz.
17. Rob Waugh, "How Did a British Polytechnic Graduate Become the Design Genius behind £200 Billion Apple?" *Mail Online*, March 20, 2011, www.dailymail.co.uk/home/moslive/article-1367481/Apples-Jonathan-Ive-How-did-British-polytechnic-graduate-design-genius.html#ixzz1PxPQwBMv.

18. "Who Is Jonathan Ive?" *Bloomberg BusinessWeek*, September 25, 2006, www.businessweek.com/magazine/content/06_39/b4002414.htm.
19. Lawrence French, "An Interview with John Lasseter, Part 3," www.fortunecity.com/skyscraper/pointone/581/interview3.html.
20. Tom Hormby, "The Pixar Story: Dick Shoup, Alex Schure, George Lucas, Steve Jobs, and Disney," January 23, 2006, http://lowendmac.com/orchard/06/pixar-story-lucas-disney.html.
21. Rebecca Keegan, "Animated—And Driven," *Los Angeles Times*, June 19, 2011, http://articles.latimes.com/2011/jun/19/entertainment/la-ca-pixar-20110619.
22. Charlie Rose, "A Discussion with Steve Jobs and John Lasseter," October 20, 1996, video, www.charlierose.com/view/interview/5885.
23. "Virgin America Upgrades 'Red' IFE System with New Features, "*Airlines and Destinations,* July 22, 2010, www.airlinesanddestinations.com/airlines/virgin-america-upgrades-red-ife-system-with-new-features.
24. Lori Zimmer, "Virgin America's New Fuel-Efficient LEAP Engines Will Save $1.6 Million per Plane," *Inhabitat,* June 16, 2011, http://inhabitat.com/virgin-america-new-fuel-efficient-leap-engines-will-save-1-6-million-per-plane.
25. Pawel Piejko, "Virgin America and Google Offer Chromebook 'Test Flights,'" *Gizmag*, July 3, 2011, www.gizmag.com/google-chromebooks-virgin-america/19089.

Chapter 7

1. GEICO Brostache Mobile App, www.geico.com/about/mobile-apps/brostache.
2. Daniel Eran Dilger, "Inside Apple's Shareholder Meeting and Q&A with Tim Cook," *Apple Insider,* February 23, 2011, www.appleinsider.com/articles/11/02/23/tim_cook_presides_over_annual_apple_shareholder_meeting.html.
3. Matt Rosoff, "Patent Troll Lodsys Says It Will Give App Store Developers $1,000 If Its Lawsuit Against Them Is Wrong," *Business Insider,* May 31, 2011, www.businessinsider.com/apple-app-store-patent-fight-heats-up-2011-5.
4. "AS/400," IBM archives, www-03.ibm.com/ibm/history/exhibits/vintage/vintage_4506VV1004.html.
5. Apple iPhone App Store web page, www.apple.com/iphone/features/app-store.html.

6. Peter Burrows, "How Apple Feeds Its Army of App Makers," *Bloomberg Businessweek,* June 8, 2011, www.businessweek.com/magazine/content/11_25/b4233039336374.htm.
7. Kit Eaton, "Apple Sues Amazon over 'App Store' Name," *Fast Company,* March 22, 2011, www.cnn.com/2011/TECH/web/03/22/apple.amazon.app.fast/index.html.
8. Jay Greene, "Microsoft Chasing Apple Lead with Money," *CNET News,* June 23, 2011, http://news.cnet.com/8301-10805_3-20073528-75/microsoft-chasing-apple-app-lead-with-money/#ixzz1SU0hVGgr.
9. Dean Takahashi, "Game Makers: Amazon's Android Appstore Terms Are Greedy," *Gamesbeat,* April 14, 2011, http://venturebeat.com/2011/04/14/game-makers-says-amazons-android-appstore-terms-are-greedy.
10. Ryan Kim, "Fred Wilson to Devs: Expect Platform Owners to Work against You," *GigaOM,* June 7, 2011, http://gigaom.com/2011/06/07/fred-wilson-to-devs-expect-platform-owners-to-work-against-you.
11. Daniel Lyons, "The Tense 'Friendship' between Facebook and Its Biggest Game Developer," *Daily Beast*, October 8, 2010, www.newsweek.com/2010/10/08/the-tense-friendship-between-facebook-and-zynga.html.
12. Mark Milian, "Some Game Developers Unhappy with Apple, Nintendo, CNN," March 4, 2011, http://edition.cnn.com/2011/TECH/gaming.gadgets/03/04/nintendo.apple.games/?hpt=Sbin.
13. USA.GOV mobile apps web page, http://apps.usa.gov.
14. John Paczkowski, "BlackBerry PlayBook: Fail with Consumers, Fail with Enterprise," *All Things D,* April 19, 2011, http://allthingsd.com/20110419/blackberry-playbook-fail-with-consumers-fail-with-enterprise.
15. Jonathan S. Geller, "Inside RIM: An Exclusive Look at the Rise and Fall of the Company that Made Smartphones Smart," *BGR*, July 13, 2011, www.bgr.com/2011/07/13/rims-inside-story-an-exclusive-look-at-the-rise-and-fall-of-the-company-that-made-smartphones-smart.
16. Ryan Tate, "London Riots Were Powered by BlackBerry," *Gawker.com*, August 8, 2011, http://gawker.com/5828800/london-riots-were-powered-by-blackberry.
17. Mike Kirkup, "What Is a Super App?" *Inside BlackBerry* (developer's blog), February 16, 2010, http://devblog.blackberry.com/2010/02/what-is-a-super-app.
18. Ian Hardy, "RIM: 'There are Over 20,000 BlackBerry Apps Available Today," *MobileSyrup*, February 14, 2011, http://mobilesyrup.com/2011/02/14/rim-there-are-over-20000-blackberry-apps-available-today.
19. Natasha Lomas, "RIM on BlackBerry Super Apps, the Apple Battle and that New OS," *Silicon.com*, July 1, 2010, www.silicon.com/technology/

mobile/2010/07/01/rim-on-blackberry-super-apps-the-apple-battle-and-that-new-os-39746044/.

20. Evan MacDonald, "Top 10 Android Apps Account for 43 Percent of App Usage," *ConsumerReports.org*, August 18, 2011, http://news.consumerreports.org/electronics/2011/08/top-10-android-apps-account-for-43-percent-of-app-usage.html.
21. Milian, "Some Game Developers Unhappy."
22. Robert Dutt, "The HP TouchPad Needs Developers, not Russell Brand," *PC World,* July 9, 2011, http://forums.precentral.net/hp-webos-news/287794-hp-touchpad-needs-developers-not-russell-brand.html.
23. Steve Jobs, "Thoughts on Flash," Apple website, April 2010, www.apple.com/hotnews/thoughts-on-flash.
24. "Blackberry Internet | RIM Demos, Native Email, Android Apps for Play-Book," *Digital Home,* May 9, 2011, http://iphoneitouch.net/Digital_Home/blackberry-internet-rim-demos-native-email-android-apps-for-playbook.
25. Jason Snell, "Jobs Speaks! The Complete Transcript," *Macworld,* October 18, 2010, www.macworld.com/article/154980/2010/10/jobs_transcript.html.

Chapter 8

1. Doug Belden profile, Hillsborough County Tax Collector web page, www.hillstax.org/AboutUs/Senior%20Staff/belden_profile.htm.
2. Cognizant web page, http://investors.cognizant.com.
3. Amy Lee, "Apple's iCloud, iTunes Match: How Apple's Music Services Measure Up," *Huffington Post,* June 6, 2011, www.huffingtonpost.com/2011/06/06/apple-icloud-itunes-match-wwdc-2011_n_872001.html.
4. "iCloud," Apple website, www.apple.com/icloud/features.
5. Aden Hepburn, "Facebook Statistics, Stats & Facts for 2011," *Digital Buzz Blog*, January 18, 2011, www.digitalbuzzblog.com/facebook-statistics-stats-facts-2011.
6. Statistics Page, Facebook website, www.facebook.com/press/info.php?statistics.
7. Rich Miller, "Facebook Unveils Custom Servers, Facility Design," *Data Center Knowledge,* April 7, 2011, www.datacenterknowledge.com/archives/2011/04/07/facebook-unveils-custom-servers-facility-design.
8. James Hamilton, "Open Compute UPS and Power Supply," *Perspectives* (blog), May 4, 2011, http://perspectives.mvdirona.com/2011/05/04/OpenComputeUPSPowerSupply.aspx.

9. James Hamilton, "Open Compute Project," *Perspectives* (blog), April 7, 2011, http://perspectives.mvdirona.com/2011/04/07/OpenComputeProject.aspx.

10. Facebook, "Facebook Launches Open Compute Project," press release, www.facebook.com/press/releases.php?p=214173.

11. Chloe Albanesius, "Greenpeace Attacks Facebook on Coal-Powered Data Center," *PC Magazine,* September 17, 2010, www.pcmag.com/article2/0,2817,2369306,00.asp.

Chapter 9

1. National Science Board, "Science and Engineering Indicators: 2010," www.nsf.gov/statistics/seind10/c0/c0i.htm.

2. Yuqing Xing and Neal Deter, "How the iPhone Widens the United States Trade Deficit with the People's Republic of China," ADBI Working Paper, December 2010, www.adbi.org/working-paper/2010/12/14/4236.iphone.widens.us.trade.deficit.prc/how.iphones.are.produced.

3. *CHaINA Magazine* web page, www.supplychain.cn/chainamag/.

4. Kurt Bauer, "Is There a Market for a Container Train China–Western Europe?" *Railway Market* CEE No.1 2008, www.railwaymarket.eu/rm2008/pdf/Is_there_a_Market_for_a_container_train_Bejing_-_Hamburg.pdf.

5. Bill Powell, "The End of Cheap Labor in China," *Time,* June 26, 2011, www.time.com/time/magazine/article/0,9171,2078121,00.html#ixzz1Pr2QF71n.

6. Frederik Balfour, "IPad Assembler Foxconn Says It Has More than 1 Million Employees in China," *Bloomberg,* December 10, 2010, www.bloomberg.com/news/2010-12-10/foxconn-says-its-china-workforce-exceeds-1-million-employees.html.

7. "Foxconn 'Mulls $12bn Brazil Move' as It Seeks Expansion," *BBC News Business,* April 13, 2011, www.bbc.co.uk/news/business-13058866.

8. "Foxconn to Replace Workers with 1 Million Robots in 3 Years," *English.News.cn,* July 30, 2011, http://news.xinhuanet.com/english2010/china/2011-07/30/c_131018764.htm.

9. Keith Bradsher, "Pearls, Finer but Still Cheap, Flow from China," *New York Times,* August 1, 2011, www.nytimes.com/2011/08/02/business/global/chinas-high-quality-pearls-enter-the-mass-market.html?nl=todaysheadlines&emc=tha2.

10. Tayfun King, "Brazil's Bid for Tech-Powered Economy," *BBC Click,* October 2, 2009, http://news.bbc.co.uk/2/hi/programmes/click_online/8284704.stm.

11. "Russian High-Tech Development Institutions Open Office in Silicon Valley," Skolkovo Foundation, March 23, 2011, http://81.177.146.232/en/newslist/20110321-silval.

12. Brett Forrest, "The Next Silicon Valley: Siberia," *Fortune,* March 26, 2007, http://money.cnn.com/magazines/fortune/fortune_archive/2007/04/02/8403482/index2.htm.

13. Andrew Rassweiler, "Japanese Earthquake Poses Potential Supply Problems for iPad2," *iSuppli*, March 17, 2011, www.isuppli.com/Teardowns/News/Pages/Japanese-Earthquake-Poses-Potential-Supply-Problems-for-iPad-2.aspx.

14. Penny Jones, "Google's Finland Data Center Pioneers New Seawater Cooling," *Data Center Dynamics,* May 25, 2011, www.datacenterdynamics.com/focus/themes/cooling/googles-finland-data-center-pioneers-new-seawater-cooling.

15. Rich Miller, "Google's Chiller-Less Data Center," *Data Center Knowledge,* July 15, 2009, www.datacenterknowledge.com/archives/2009/07/15/googles-chiller-less-data-center.

16. Rich Miller, "Dublin Emerges as Cloud Computing Hub," *Data Center Knowledge,* March 17, 2011, www.datacenterknowledge.com/archives/2011/03/17/dublin-emerges-as-cloud-computing-hub.

17. Rich Miller, "Dell Plans Global Network of Cloud Data Centers," www.datacenterknowledge.com/archives/2011/04/07/dell-plans-global-network-of-cloud-data-centers/.

18. "Singapore Emerging as the Cloud Computing Hub in Asia-Pacific," *IT Times,* September 10, 2010, www.koreaittimes.com/story/10397/singapore-emerging-cloud-computing-hub-asia-pacific.

19. "The Broadband Economy," Intelligent Community Forum, 2011, www.intelligentcommunity.org/index.php?src=gendocs&ref=TheBroadbandEconomy&category=AboutUs.

20. Kyle Peterson, "Special Report: A Wing and a Prayer: Outsourcing at Boeing," *Reuters*, January 20, 2011, www.reuters.com/article/2011/01/20/us-boeing-dreamliner-idUSTRE70J2UX20110120?pageNumber=5.

21. Jeff Moad, "Boeing's Big Supply Chain Wager," *Managing Automation,* July 23, 2007, www.managingautomation.com/maonline/magazine/read/view/Boeing__39_s_Big_Supply_Chain_Wager_17760261.

22. Boeing, "First Major Assembly for Boeing 787 Dreamliner Delivered to Everett," press release, April 25, 2007, www.boeing.com/news/releases/2007/q2/070425b_pr.html.

23. Jason Paur, "Boeing's 787, as Innovative Inside as Outside," *CNN Tech*, December 25, 2009, http://articles.cnn.com/2009-12-25/tech/wired.boeing.787.inside_1_dreamliner-cabin-light?_s=PM:TECH.

24. Marilyn Adams, "Dreamliner Designed so Fliers Can Breathe Easy," *USA Today,* November 1, 2006, www.usatoday.com/money/biztravel/2006-10-30-boeing-air-usat_x.htm.
25. Stanley Holmes, "Boeing's Plastic Dream Machine," *Bloomberg Businessweek,* June 20, 2005, www.businessweek.com/magazine/content/05_25/b3938037_mz011.htm.
26. Peterson, "A Wing and a Prayer."
27. Erik Sofge, "Boeing's New 787 Dreamliner: How It Works," *Popular Mechanics,* August 4, 2006, www.popularmechanics.com/technology/aviation/news/boeing-787.

Chapter 10

1. "Outback Steakhouse, Inc.," profile, Funding Universe, www.fundinguniverse.com/company-histories/Outback-Steakhouse-Inc-Company-History.html.
2. Chris Murphy, "Procter & Gamble CIO Filippo Passerini: 2010 Chief of the Year," *InformationWeek,* December 10, 2010, www.informationweek.com/news/global-cio/interviews/228500182.
3. Jeff Dyer, Hal Gregersen, and Clayton Christersen, "The DNA—People, Processes, and Philosophies—of Innovative Companies," *Fast Company,* July 21, 2011, www.fastcompany.com/article/innovators-dna-clayton-christensen-jeff-dyer-hal-gregersen.
4. Vinnie Mirchandani, *The New Polymath* (Hoboken, NJ: John Wiley & Sons, 2010).
5. Vineet Nayyar, *Employees First, Customers Second: Turning Conventional Management Upside Down* (Cambridge, MA: Harvard Business Press, 2010).
6. William C. Taylor and Polly G. Labarre, *Mavericks at Work: Why the Most Original Minds in Business Win* (New York: William Morrow, 2006).
7. "50 Maverick Minds," www.mavericksatwork.com.
8. Warren Brown, "Ugly Has Never Been so Attractive," *Washington Post,* August 10, 2008, www.washingtonpost.com/wp-dyn/content/article/2008/08/08/AR2008080801388.html.
9. Tom Simonite, "A Tablet that Wants to Take Over the Desktop," *MIT Technology Review,* June 29, 2011, www.technologyreview.com/computing/37936/?p1=A4&a=f.
10. Joshua Topolsky, "Motorola Atrix 4G Review," *Engadget,* June 29, 2011, www.engadget.com/motorola/atrix-4g-review.
11. "iPad Smart Cover," Apple web page, www.apple.com/ipad/smart-cover.

12. Brad Smith, "Watson Developer Speaks Out against Apple; Plans Port to Windows," *MacObserver,* July 28, 2002, www.macobserver.com/tmo/article/Watson_Developer_Speaks_Out_Against_Apple_Plans_Port_To_Windows.

13. Dan Wood, "The Long Story behind Karelia's New Logo," *Karelia Blog,* January 6, 2006, www.karelia.com/news/small_and_nimble_the_long_s.html.

14. Daniel Eran Dilger, "Inside Apple's Shareholder Meeting and Q&A with Tim Cook," *Apple Insider,* February 23, 2011, www.appleinsider.com/articles/11/02/23/tim_cook_presides_over_annual_apple_shareholder_meeting.html.

15. Daniel Eran Dilger, "Apple iOS App Store Blamed for Too Many Apps as Sony NGP Is Called 'Dead on Arrival,'" *Apple Insider,* March 4, 2011, www.appleinsider.com/articles/11/03/04/apple_ios_app_store_blamed_for_too_many_apps_as_sony_ngp_is_called_dead_on_arrival.html.

16. Matt Rosoff, "Apple Steps in to Crush Patent Troll," *BusinessInsider,* June 10, 2011, www.businessinsider.com/apple-gets-serious-about-taking-on-patent-troll-2011-6?op=1#ixzz1ST0MZ0El.

17. Peter Burrows, "The Seed of Apple's Innovation," *Bloomberg Businessweek,* October 12, 2004, www.businessweek.com/bwdaily/dnflash/oct2004/nf20041012_4018_db083.htm.

18. Austin Carr, "The State of Internet Music on YouTube, Pandora, iTunes, and Facebook," *Fast Company,* July 20, 2010, www.fastcompany.com/1672447/the-state-of-internet-music-on-youtube-pandora-itunes-and-facebook.

19. Betsy Morris, "Steve Jobs Speaks Out," *Fortune,* August 3, 2008, http://money.cnn.com/galleries/2008/fortune/0803/gallery.jobsqna.fortune/13.html.

20. Paul Dunay, "Clayton Christensen Said the iPhone Will Not Succeed," Buzz Marketing for Technology, June 15, 2007, http://pauldunay.com/clayton-christensen-said-iphone-will.

21. Marc Flores, "RIM Called Apple Big Ol' Fakers when iPhone Was Unveiled in 2007," *IntoMobile,* December 27, 2010, www.intomobile.com/2010/12/27/rim-apple-iphone-unveiled-2007.

22. Will Park, "Activate Your Apple iPhone from Home Using iTunes!" *IntoMobile,* June 26, 2007, www.intomobile.com/2007/06/26/activate-your-apple-iphone-from-home-using-itunes.

23. Mig Siegler, "AT&T Is a Big, Steaming Heap of Failure," *TechCrunch,* July 18, 2009, http://techcrunch.com/2009/07/18/att-is-a-big-steaming-heap-of-failure.

24. Joel Johnson, "1 Million Workers. 90 Million iPhones. 17 Suicides. Who's to Blame?" *Wired,* March 2011, www.wired.com/magazine/2011/02/ff_joelinchina/all/1.

25. Cliff Edwards, "Commentary: Sorry, Steve: Here's Why Apple Stores Won't Work," *Bloomberg Businessweek,* May 21, 2001, www.businessweek .com/magazine/content/01_21/b3733059.htm.

26. Joe Wilcox, "Does Apple Have a Future in Retail?" *C/Net*, May 15, 2001, http://news.cnet.com/Does-Apple-have-a-future-in-retail/2100-1040_3-257629.html?tag=mncol#ixzz1SQ6RnzZj.

27. Harry McCracken, "The Long Fail: A Brief History of Unsuccessful Tablet Computers," *Technologizer,* January 27, 2010, http://technologizer.com/2010/01/27/the-long-fail-a-brief-history-of-unsuccessful-tablet-computers/.

28. Vinnie Mirchandani, *The New Polymath* (Hoboken, NJ: John Wiley & Sons, 2010).

29. Catherine Smith, "Dell Says Apple iPad Will Fail in Enterprise," *Huffington Post,* March 30, 2011, www.huffingtonpost.com/2011/03/30/dell-apple -ipad-will-fail-too-expensive_n_842480.html.

30. Brian Barrett, "Steve Jobs Gets Candid on iPad, Kindle, and Love Battery Life," *Gizmodo*, January 29, 2010, http://gizmodo.com/5459576/steve-jobs -gets-candid-on-ipad-kindle-and-love-battery-life.

31. Tim Nudd, "Apple's 'Get a Mac,' the Complete Campaign," *AdWeek,* April 13, 2011, www.adweek.com/adfreak/apples-get-mac-complete-campaign -130552.

32. Steve Jobs, "Thoughts on Flash," Apple website, April 2010, www.apple .com/hotnews/thoughts-on-flash.

33. Arnold Kim, "Apple Announces Long-Term Flash Memory Agreements," *MacRumors,* November 21, 2005, www.macrumors.com/2005/11/21/apple -announces-long-term-flash-memory-agreements.

34. Brian X. Chen, "4th Time a Charm for Apple? From iDisk to .Mac to MobileMe to iCloud," *Wired,* May 31, 2011, www.wired.com/epicenter/2011/05/icloud-apple.

35. Amy Lee, "Apple's iCloud, iTunes Match: How Apple's Music Services Measure Up," *Huffington Post,* June 6, 2011, www.huffingtonpost.com/2011/06/06/apple-icloud-itunes-match-wwdc-2011_n_872001.html.

36. "iCloud Features," Apple web page, www.apple.com/icloud/features.

Chapter 11

1. Jonny Evans, "iPad 2 Wins the Tablet Wars, Competitors already DOA," *Computerworld*, March 8, 2011, http://blogs.computerworld.com/17940/ipad_2_wins_the_tablet_wars_competitors_already_d_o_a.

2. Jason Perlow, "HP's TouchPad: Dead on Arrival," *ZDNet,* June 20, 2011, www.zdnet.com/blog/perlow/hps-touchpad-dead-on-arrival/17577?tag=nl.e550.

3. Renee Hopkins, "Stuck? Take a Look at Your Business Model," *Bloomberg Businessweek,* February 17, 2010, www.businessweek.com/innovate/content/feb2010/id20100217_897247_page_3.htm.

4. Ken Auletta, "Publish or Perish," *The New Yorker,* April 26, 2010, www.newyorker.com/reporting/2010/04/26/100426fa_fact_auletta#ixzz1PkpTkCyX.

5. Motoko Rich and Brad Stone, "Amazon Threatens Publishers as Apple Looms," *New York Times,* March 17, 2010, www.nytimes.com/2010/03/18/technology/internet/18amazon.html.

6. Paid advertising message from Macmillan, Publishers Lunch, January 30, 2010, www.publishersmarketplace.com/lunch/macmillan_30jan10.html.

7. Boston Consulting Group, "Business Model Innovation: When the Going Gets Tough, Change the Game," www.bcg.com/documents/file36456.pdf.

8. Nick Clayton, "Rebooting Business," *Wall Street Journal,* June 29, 2011, http://online.wsj.com/article/SB10001424052702303714704576382871251873828.html?mod=WSJEUROPE_hps_MIDDLEFourthNews.

9. Fred Vogelstein, "The Untold Story: How the iPhone Blew Up the Wireless Industry," *Wired*, February 2008, www.wired.com/gadgets/wireless/magazine/16-02/ff_iphone?currentPage=4.

10. Ben Popken, "Despite No Service, AT&T Refuses to Waive ETF," *The Consumerist*, May 17, 2011, http://consumerist.com/2011/05/despite-no-service-att-refuses-to-waive-etf.html.

11. Mike Cottmeyer, "Watch Out for Excessive AT&T/iPhone Roaming Charges," *LeadingAgile* blog, November 21, 2008, www.leadingagile.com/?s=roaming.

12. "Consumer Reports Cell-Service Ratings: AT&T Is the Worst Carrier," *ConsumerReports.org*, December 6, 2010, http://news.consumerreports.org/electronics/2010/12/consumer-reports-cell-phone-survey-att-worst.html.

13. Brian X. Chen, "Verizon iPhone Shows You Can't Win: Carriers Hold the Cards," *Wired*, February 7, 2011, www.wired.com/gadgetlab/2011/02/iphone-verizon-sucks.

14. David Meyer, "Android Shoots Past iPhone OS in Market Share," *ZDNet UK*, May 19, 2011, www.zdnet.co.uk/news/mobile-devices/2011/05/19/android-shoots-past-iphone-os-in-market-share-40092829.

15. Kevin C. Tofel, "The Future of Cheap Androids Begins Now," *GigaOM*, February 2, 2011, http://gigaom.com/mobile/cheap-android-smartphones.

16. Apache Hadoop overview, http://hadoop.apache.org.
17. Mark Guarino, "Three Ways iTunes, and Its 10 Billion in Sales, Changed Music Industry," *Christian Science Monitor,* February 26, 2010, www.csmonitor.com/USA/Society/2010/0226/Three-ways-iTunes-and-its-10-billion-in-sales-changed-music-industry.
18. "EMI Music Launches DRM-Free iTunes Downloads in Higher-Quality," *AppleInsider,* April 2, 2007, www.appleinsider.com/articles/07/04/02/emi_music_launches_drm_free_itunes_downloads_in_higher_quality.html.
19. Amy Lee, "Apple's iCloud, iTunes Match: How Apple's Music Services Measure Up," *Huffington Post,* June 6, 2011, www.huffingtonpost.com/2011/06/06/apple-icloud-itunes-match-wwdc-2011_n_872001.html.
20. "iCloud Features," Apple web page, www.apple.com/icloud/features.
21. Collin Smith, "Apple's New iCloud: Legalizing Illegal Music?" *Next Gen Journal*, June 7, 2011, http://nextgenjournal.com/2011/06/apples-new-icloud-boosts-consumer-convenience-and-music-industry-profits.
22. "Government and IT—A Recipe for Rip-Offs: Time for a New Approach," Public Administration Committee of the UK Parliament, www.publications.parliament.uk/pa/cm201012/cmselect/cmpubadm/715/71504.htm.
23. "U.S. Tops World in Health Spending, Results Lag," Associated Press, December 8, 2009, www.msnbc.msn.com/id/34330376/ns/health%20-%20health_care.
24. U.S. Department of Health and Human Services, "The New Numbers—Health Insurance Reform Cannot Wait," news release, September 16, 2009, www.hhs.gov/news/press/2009pres/09/20090916b.html.
25. Timothy K. Lake, Kate A. Stewart, and Paul B. Ginsburg, "Lessons from the Field: Making Accountable Care Organizations Real," National Institute for Health Care Reform, January 20, 2011, www.nihcr.org/Accountable-Care-Organizations.html.
26. Phil Galewitz, "ACOs: A Quick Primer," *Kaiser Health News,* July 17, 2009, www.kaiserhealthnews.org/Stories/2009/July/17/ACO.aspx.

Chapter 12

1. Cliff Edwards, "Commentary: Sorry, Steve: Here's Why Apple Stores Won't Work," *Bloomberg BusinessWeek,* May 21, 2001, www.businessweek.com/magazine/content/01_21/b3733059.htm.
2. Myriam Joire, "GE Kicks Off EV Experience Tour, Promises WattStations for All," *Engadget,* March 14, 2011, www.engadget.com/2011/03/14/ge-kicks-off-ev-experience-tour-promises-wattstations-for-all.

3. Penny Crosman, "Citibank Opens Apple Store-Like Branch in Union Square," *Bank Systems and Technology*, December 17, 2010, www.banktech.com/channels/228800795.

4. Tim Greene, "Al Gore and Cisco's John Chambers Get Green Together," *Network World*, March 19, 2008, www.networkworld.com/news/2008/031908-voicecon-gore-chambers-green.html.

5. Cisco Communities, blog post on VoiceCon, Spring 2008, January 23, 2009, https://communities.cisco.com/docs/DOC-2320.

6. Robert Spector, *Category Killers: The Retail Revolution and Its Impact on Consumer Culture* (Cambridge, MA: Harvard Business Press, 2005).

7. Anita Hamilton, "Why Circuit City Busted, while Best Buy Boomed," *Time,* November 11, 2008, www.time.com/time/business/article/0,8599,1858079,00.html#ixzz1OamDBWc1.

8. Eric Bangeman, "CompUSA Closing over Half of Its Stores," *ARS Technica,* February 28, 2007, http://arstechnica.com/old/content/2007/02/8940.ars.

9. Claer Barrett, "Electrical Stores Ramp up Customer Service," *FT.com*, May 30, 2011, www.ft.com/intl/cms/s/0/dd889086-8afc-11e0-b2f1-00144feab49a.html#axzz1Ob0zRDd2.

10. May Wong, "Best Buy Service Trumps Circuit City," *Washington Post,* April 6, 2007, www.washingtonpost.com/wp-dyn/content/article/2007/04/06/AR2007040601194_3.html.

11. Doug Williams, "Case Study: Best Buy's Retailer-Led Product Strategy," Forrester Research, June 30, 2010, www.forrester.com/rb/Research/case_study_best_buys_retailer-led_product_strategy/q/id/57080/t/2.

12. Vinnie Mirchandani, *The New Polymath* (Hoboken, NJ: John Wiley & Sons, 2010).

13. Diann Daniel, "Geeks at Your Service: Secret of Best Buy's Success," *Network World*, November 13, 2008, www.networkworld.com/news/2008/111308-geeks-at-your-service-secret.html.

14. Claire Cain Miller, "Best Buy Backs a Digital Media Venture Fund," *New York Times,* May 27, 2009, http://bits.blogs.nytimes.com/2009/05/27/best-buy-backs-a-digital-media-venture-fund.

15. Thomas Yarnell, "Best Buy's Twelpforce: A Social Media Success Story? (in progress)," Mirsky and Company, *PLLC* blog, August 6, 2010, http://mirskylegal.com/2010/08/best-buy%E2%80%99s-twelpforce-a-social-media-success-story-in-progress.

16. Steven Russolillo, "Best Buy 4Q Exceeds Expectations, But Challenges Persist," *Fox Business*, March 24, 2011, www.foxbusiness.com/industries/2011/03/24/best-buy-4q-exceeds-expectations-challenges-persist/#ixzz1ObusElvk.

17. Alan Wolf, "Best Buy Details Strategic Growth Plans," *Twice.com*, April 18, 2011, www.twice.com/article/466828-Best_Buy_Details_Strategic_Growth_Plans.php.

18. Associated Press, "Best Buy Rolling Out Smaller Stores," March 24, 2011, http://jacksonville.com/business/2011-03-24/story/best-buy-rolling-out-smaller-stores#ixzz1Oc1uom2R.

19. Michael Gartenberg, "Apple Keeps on Giving Lessons in Retail," *Macworld,* May 22, 2011, www.macworld.com/article/159990/2011/05/gartenberg_lessons_in_apple_retail.html.

20. Josh Lowensohn, "Microsoft to Open Third Retail Store in California," *C/Net*, January 19, 2011, http://news.cnet.com/8301-10805_3-20028944-75.html.

21. Vincent Chang, "Nokia Closes Three More Flagship Retail Stores," *Cellular-News*, December 10, 2009, www.cellular-news.com/story/41025.php.

22. "Frequently Asked Questions," Walgreens web page, http://news.walgreens.com/article_display.cfm?article_id=831#10.

23. Chris Burritt, "Home Depot's Fix-It Lady," *Bloomberg Businessweek,* January 13, 2011, www.businessweek.com/magazine/content/11_04/b4212064704660.htm.

24. "Surface Computer Arrives at AT&T First," *Forbes,* April 2, 2008, www.forbes.com/feeds/afx/2008/04/02/afx4846714.html.

25. "The World's Most Innovative Companies 2011," *Fast Company*, March 2011, www.fastcompany.com/most-innovative-companies/2011.

26. Martyn Williams, "Adidas Previews Interactive Shopping Wall," *PC World*, March 3, 2011, www.pcworld.com/businesscenter/article/221290/adidas_previews_interactive_shopping_wall.html.

27. Peter Fingar, *Extreme Competition: Innovation and the Great 21st Century Business Reformation* (Meghan Kiffer Press, 2006).

28. "Kate and Pippa Middleton's Post-Wedding Looks: All the Details!" *People Magazine*, May 3, 2011, http://stylenews.peoplestylewatch.com/2011/05/03/kate-middleton-pippa-zara.

29. Thomas K. Grose, "Not too Posh to Click," *Time Magazine*, June 27, 2011, www.time.com/time/magazine/article/0,9171,2078093,00.html#ixzz1Pr57pgPG.

30. "Overview," Taubman web page, www.taubman.com/abouttaubman/about_overview.html.

31. "Taubman Centers/Sharp Electronics Corporation Alliance Creates High Definition Experience for Shoppers," press release, *PR Newswire*, May 21, 2007, www.prnewswire.com/news-releases/taubman-centerssharp

-electronics-corporation-alliance-creates-high-definition-experience-for-shoppers-58358707.html.

32. John Jannarone, "Apple Upsets the Department-Store Cart," *Wall Street Journal,* June 10, 2011, http://online.wsj.com/article/SB10001424052702304259304576375833098136732.html.

Chapter 13

1. Dan Nosowitz, "The History of the Teardown: The Need to See Our Gear Undressed," *Popsci,* March 29, 2011, www.popsci.com/node/52862/?cmpid=enews033111.
2. Andrew Bookholt, "Installing iPhone 4 Verizon 5-Point Pentalobe Screws," *iFixit,* www.ifixit.com/Guide/Repair/Installing-iPhone-4-Verizon-5-Point-Pentalobe-Screws/4880/1.
3. "IPhone Jailbreak History," *Quali-tv.com,* March 31, 2011, www.quali-tv.com/news/general/iphone-jailbreak-history.
4. Derek Scott, "Rooting for Dummies: A Beginner's Guide to Rooting Your Android Device," *Android Authority,* March 22, 2011, www.androidauthority.com/rooting-for-dummies-a-beginners-guide-to-root-your-android-phone-or-tablet-10915.
5. Duncan Rowe Graham, "Sony Sues over PS3 Encryption Hack," *NewScientist,* January 14, 2011, www.newscientist.com/article/dn19973-sony-sues-over-ps3-encryption-hack.html.
6. James E-Smith, "PS3 Hacking Fallout Continues," *My Gaming,* January 19, 2011, http://mygaming.co.za/news/ps3/9465-PS3-hacking-fallout-continues.html.
7. Jason Mick, "Editorial: LulzSec Targets Elderly in the Wake of Latest Sony Hacks," *Daily Tech,* June 5, 2011, www.dailytech.com/EDITORIAL+LulzSec+Targets+Elderly+in+the+Wake+of+Latest+Sony+Hacks/article21819.htm.
8. Bill Saporito, "Hack Attack," *Time Business,* June 23, 2011, www.time.com/time/business/article/0,8599,2079423,00.html#ixzz1QPwmJADh.
9. "Professional Security Events," Black Hat, www.blackhat.com/html/about.html.
10. "2011 CWE/SANS Top 25 Most Dangerous Software Errors," Common Weakness Enumeration, September 13, 2011, http://cwe.mitre.org/top25/archive/2011/2011_cwe_sans_top25.pdf.

11. Mike Wehner, "5 Things You Probably Didn't Know Could Be Hacked," *Yahoo!*, August 15, 2011, http://news.yahoo.com/blogs/technology-blog/5-things-probably-didn-t-know-could-hacked-174330493.html.
12. "War Texting: Hackers Unlock Car Doors via SMS," *The Hacker News*, July 28, 2011, www.thehackernews.com/2011/07/war-texting-hackers-unlock-car-doors.html.
13. Mike Nathan, "Home Automation Systems Easily Hacked via the Power Grid," *Hack a Day,* August 8, 2011, http://hackaday.com/2011/08/08/home-automation-systems-easily-hacked-via-the-power-grid.
14. Jason Snell, "Inside Apple's Once-Secret Wireless Lab," *Macworld*, July 16, 2010, www.macworld.com/article/152771/2010/07/wireless_lab.html.
15. "Security First: Security and Data Protection in Google Data Centers," *Google Enterprise Blog*, April 22, 2011, http://googleenterprise.blogspot.com/2011/04/security-first-security-and-data.html.
16. Leslie Horn, "Apple's Billion-Dollar Data Center Suddenly Appears on Google Earth," *PCMAG*, June 1, 2011, www.pcmag.com/article2/0,2817,2386259,00.asp.
17. "A New Approach to China," *The Official Google Blog*, January 12, 2010, http://googleblog.blogspot.com/2010/01/new-approach-to-china.html.
18. Ashlee Vance, "Welcome to Las Vegas—Home of the Technology Superpower You've Never Heard of," *The Register*, May 24, 2008, www.theregister.co.uk/2008/05/24/switch_switchnap_rob_roy.
19. Don Clark, "Vegas Computer Center Locked Down, Always Up," *Wall Street Journal,* July 14, 2011, http://blogs.wsj.com/digits/2011/07/14/vegas-computer-center-locked-down-always-up.
20. Jesus Diaz, "How Apple Lost the Next iPhone," *Gizmodo*, April 19, 2010, http://gizmodo.com/5520438/how-apple-lost-the-next-iphone.
21. Mark L. Krotoski, "Common Issues and Challenges in Prosecuting Trade Secret and Economic Espionage Cases," *United States Attorneys' Bulletin*, November 2009, www.justice.gov/usao/eousa/foia_reading_room/usab5705.pdf.

Chapter 14

1. Owen O'Mally, "I'll Take Hadoop for $400, Alex," *Yahoo!*, Feburary 24, 2011, http://developer.yahoo.com/blogs/hadoop/posts/2011/02/i%E2%80%99ll-take-hadoop-for-400-alex.
2. "Final Jeopardy Goes to... Hadoop," *Yahoo!*, February 18, 2011, http://ycorpblog.com/2011/02/18/jeopardy-hadoop/.

3. Derrick Harris, "As Big Data Takes Off, the Hadoop Wars Begin," *GigaOM*, March 25, 2011, http://gigaom.com/cloud/as-big-data-takes-off-the-hadoop-wars-begin.
4. "The Open Kinect Project—THE OK PRIZE—Get $3,000 Bounty for Kinect for Xbox 360 Open Source Drivers," *Adafruit* blog, November 4, 2010, www.adafruit.com/blog/2010/11/04/the-open-kinect-project-the-ok-prize-get-1000-bounty-for-kinect-for-xbox-360-open-source-drivers.
5. Tor Thorsen, "Microsoft Denies Kinect Hack Claims," *Gamespot UK*, November 8, 2010, http://uk.gamespot.com/xbox360/sports/thebiggestloser ultimateworkout/news.html?sid=6283696&tag=mantle_skin;content.
6. Mike Schramm, "Kinect: The Company behind the Tech Explains How It Works," *Joystiq*, June 19, 2010, www.joystiq.com/2010/06/19/kinect-how-it-works-from-the-company-behind-the-tech.
7. Dana Blankenhorn, "Microsoft Surrenders on Linux Kinect Hack," *ZDNet*, November 12, 2010, www.zdnet.com/blog/open-source/microsoft-surrenders-on-linux-kinect-hack/7769.
8. "Introducing OpenNI," OpenNI, http://openni.org/about.
9. Desire Athow, "Microsoft's Lawsuit Payout Amounts to around $9 Billion," *The Inquirer*, July 14, 2005, www.theinquirer.net/inquirer/news/1048246/microsoft-lawsuit-payouts-usd9-billion#ixzz1Q6XQk2iZ.
10. "Microsoft's Lucrative New Revenue Stream? Android," *All Things D*, May 27, 2011, http://allthingsd.com/20110527/microsofts-lucrative-new-revenue-stream-android.
11. Andrew Nusca, "Oracle, Google Scrap over Android IP Damages," *ZDNet*, June 30, 2011, www.zdnet.com/blog/btl/oracle-google-scrap-over-android-ip-damages/51696.
12. "Nortel Patent Sale: How It Could Spell Trouble for Google," *International Business Times*, July 2, 2011, www.ibtimes.com/articles/173205/20110702/nortel-patent-sale-google-apple-inc-emc-corp-microsoft-corp-android.htm.
13. Susan Decker, "Google Lawyer Says Patents Are 'Gumming Up' Innovation," *Bloomberg,* July 26, 2011, http://mobile.bloomberg.com/news/2011-07-26/google-general-counsel-says-patents-are-gumming-up-innovation?category=.
14. Peter Svensson, "Google Buys about 1,000 IBM Patents," *Huffingpost Tech*, August 1, 2010, www.huffingtonpost.com/2011/08/01/google-ibm-patents_n_914858.html.
15. "Supercharging Android: Google to Acquire Motorola Mobility," *The Official Google Blog*, August 15, 2011, http://googleblog.blogspot.com/2011/08/supercharging-android-google-to-acquire.html.

16. Verizon Communications 10-K December 31, 2010 filed with the U.S. SEC, www.sec.gov/Archives/edgar/data/732712/000119312511049476/d10k.htm #tx62372_5.
17. Sam Williams, A heaven for patent pirates, *Technology Review*, February 3, 2006 available at www.technologyreview.com/communications/16280/page1/.
18. Carl Sifakis, *The Mafia Encyclopedia* (New York: Checkmark Books, 1999).

Chapter 15

1. Charlie Feld, *Blind Spot: A Leader's Guide to IT-Enabled Business Transformation* (Olive Press, 2010).
2. Charles H. Fine, *Clockspeed: Winning Industry Control in the Age of Temporary Advantage* (New York: Basic Books, 1999).
3. Sam Grobart, "For Flip Video Camera, Four Years from Hot Start-Up to Obsolete," *New York Times*, April 12, 2011, www.nytimes.com/2011/04/13/technology/13flip.html.
4. Dan Frommer, "Chart of the Day: The iPhone Is Now Half of Apple's Business," *Business Insider,* April 21, 2011, www.businessinsider.com/chart-of-the-day-apple-revenue-by-segment-2011-4#ixzz1OoHs4H3o.
5. Seth Porges, "5 New GPS Features to Guide You This Year: Live @ CES 2008," *Popular Mechanics*, October 1, 2009, www.popularmechanics.com/technology/gadgets/4243529.
6. "What We're Driving at," *The Official Google Blog*, October 9, 2010, http://googleblog.blogspot.com/2010/10/what-were-driving-at.html.
7. Ryan Beene, "Toyota, Microsoft's Team on Telematic Software, "*Autoweek,* April 6, 2011, www.autoweek.com/article/20110406/CARNEWS/110409930.
8. Olga Kharif, "Samsung, LG Take Aim at Whirlpool with Smart Appliances," *Bloomberg Businessweek*, January 26, 2011, www.businessweek.com/technology/content/jan2011/tc20110126_265727.htm.
9. UBM TechInsights, "UBM TechInsights Issues Warning to Medical Devices Industry about Possible Obsolescence as Reliance on Smartphones Increases," press release, February 7, 2011, www.ubmtechinsights.com/uploadedFiles/UBM-TechInsights-Medical-Device-Convergence-02072011.pdf.
10. Devin Coldewey, "iOS Passes 200 Million Devices, 25 Million of which Are iPads," *Techcrunch*, June 6, 2011, http://techcrunch.com/2011/06/06/ios-passes-200-million-devices-25-million-of-which-are-ipads.

11. Paul Miller, "Microsoft Announces 2.5 Million Kinects Sold in First 25 Days," *Engadget*, November 29, 2010, www.engadget.com/2010/11/29/microsoft-announces-2-5-million-kinects-sold-in-first-25-days.

12. Brad Stone, "Turning Page, e-Books Start to Take Hold," *New York Times*, December 23, 2008, www.nytimes.com/2008/12/24/technology/24kindle.html?_r=1&scp=2&sq=kindle&st=cse.

13. "An e-Reader Shortage for the Holidays?" *CNN*, November 27, 2009, http://scitech.blogs.cnn.com/2009/11/27/an-e-reader-shortage-for-the-holidays.

14. Jennifer Devries-Valentino, "Amazon's e-Reader Out of Stock: Is It Demand, or a Kindle 3?" *WSJ Blogs*, July 28, 2010, http://blogs.wsj.com/digits/2010/07/28/amazons-e-reader-out-of-stock-is-it-demand-or-a-kindle-3.

15. Thomas Wailgum, "Nintendo Wii Shortage: Shrewd Marketing or Flawed Supply Chain?" *CIO*, August 22, 2008, www.cio.com/article/445316/Nintendo_Wii_Shortage_Shrewd_Marketing_or_Flawed_Supply_Chain_.

16. Yi-Wyn Yen, "No Slowdown for Wii and DS, Says Nintendo Prez," *CNN Money*, October 6, 2008, http://techland.blogs.fortune.cnn.com/2008/10/06/no-slowdown-for-wii-and-ds-says-nintendo-prez.

17. Chris Davies, "Kindle eBook Sales Exceed Print Sales in US," *Slash Gear*, May 19, 2011,www.slashgear.com/kindle-ebook-sales-exceed-print-sales-in-us-19153084.

18. Preston Chesser, "History Articles: The Burning of the Library of Alexandria," *Ehistory*, http://ehistory.osu.edu/world/articles/ArticleView.cfm?AID=9.

19. "About the Internet Archive," www.archive.org/about/about.php.

20. Diane Bartz, "Judge Slaps Down Google's Digital Library Settlement," *Reuters*, March 22, 2011, www.reuters.com/article/2011/03/22/us-google-books-antitrust-idUSTRE72L6D920110322.

21. Davis Dyer and Daniel Gross, *The Generations of Corning: The Life and Times of a Global Corporation* (Oxford, UK: Oxford University Press, 2001).

22. "Frequently Asked Questions: Corning Gorilla Glass," www.corninggorillaglass.com/faqs/all.

23. "Corning Gorilla Glass," www.Corninggorillaglass.com.

24. Presentation on Gorilla Glass at Corning Annual Shareholder meeting, April 28, 2011, http://files.shareholder.com/downloads/glw/1143077664x0x463008/553ac025-c761-4366-9af2-9e385f0f78e0/Shareholders_2011_FINAL_SteinerPDF.pdf.

25. "Is Gorilla Glass a Big Magilla?" *HDGuru*, November 4, 2010, http://hdguru.com/is-gorilla-glass-a-big-magilla/2721.

26. www.corninggorillaglass.com/SonyBRAVIA.

27. "Hyundai Blue2," *ZerCustoms*, March 31, 2011, www.zercustoms.com/news/Hyundai-Blue2.html.

28. Ariel Schwartz, "Soon We May All Live in Prefab High-Rises," *Fast Company,* April 14, 2011, www.fastcompany.com/1747150/soon-we-will-all-live-in-prefab-high-rise-buildings.

Chapter 16

1. Guy Kawasaki, *Enchantment: The Art of Changing Hearts, Minds, and Actions* (New York: Portfolio, 2011).

2. Guy Kawasaki, "HOW TO: Launch Any Product Using Social Media," *Mashable,* March 30, 2011, http://mashable.com/2011/03/30/product-launch-social-media.

3. Dennis Stoutenburg, "Measuring the Social Media Chatter of the 2011 Super Bowl Commercials: Who Were the Real Winners?" www.socialstrategy1.com/wp-content/uploads/2010/03/case_studies/wp-Social_Media_and_Super_Bowl_Advertisers.pdf.

4. "Brand Bowl 2011: Recap," *Boston.com*, http://brandbowl2011.com.

5. Maddi Hausmann Sojourner, "Google I/O 2011 Wrap-Up," *Android Headlines*, May 13, 2011, http://androidheadlines.com/2011/05/google-io-2011-wrap-up.html.

6. Ed Morita, "Review of Cirque du Soleil 'Beatles Love,'" *Nonstop Online Entertainment,* Honolulu, August 9, 2010, www.nonstophonolulu.com/entertainment/review-of-cirque-du-soleil-beatles-love/.

7. Bryan Lam, "Exclusive: Dell Mini Inspiron, Their First Mini Laptop," *Gizmodo*, May 28, 2008, http://gizmodo.com/393815/exclusive-dell-mini-inspiron-their-first-mini-laptop.

8. "Dell Social Media Case Study" *Radian6*, www.radian6.com/resources/library/dell.

9. Bill Brenner, "Google+ a Security Risk? I'm There!" *CSO Blogs,* July 25, 2011, http://blogs.csoonline.com/1613/google_a_security_risk_im_there.

10. Marc Andreessen, "Why Software Is Eating the World," *Wall Street Journal*, August 20, 2011, http://online.wsj.com/article/SB10001424053111903480904576512250915629460.html?mod=WSJ_hp_mostpop_read.

11. Adam Ostrow, "Exclusive: Sprint's Social Media Campaign for HTC EVO 4G Debuts Tonight," *Mashable*, June 3, 2010, http://mashable.com/2010/06/03/sprint-htc-evo-social-media.

12. "Spammers Exploit Craze for Google+ Invites," *International Business Times,* July 1, 2011, www.ibtimes.com/articles/173149/20110701/google-invite.htm.
13. John Quelch, "How to Profit from Scarcity," *HBS Working Knowledge*, September 14, 2007, http://hbswk.hbs.edu/item/5776.html.
14. "Salesforce.com and Toyota Form Strategic Alliance to Build 'Toyota Friend' Social Network for Toyota Customers and Their Cars," *Toyota Newsroom*, May 23, 2011, http://pressroom.toyota.com/releases/toyota+friend+social+network.htm.

Chapter 17

1. "Know the Global Warming Score," DriveClean.ca.gov, www.driveclean.ca.gov/Know_the_Score/Know_the_Global_Warming_Score.php.
2. "Scoring Cell Phones," *GoodGuide*, www.goodguide.com/categories/332304-cell-phones##btr.
3. "Prospect of Limiting the Global Increase in Temperature to 2°C Is Getting Bleaker," International Energy Agency, May 30, 2011, www.iea.org/index_info.asp?id=1959.
4. "Environmental Sustainability," HP Global Citizenship web page, www.hp.com/hpinfo/globalcitizenship/environment/environmental_sustainability.html.
5. "Apple Shuns CSR Report at February's Shareholder Meeting," *The Green Economy*, www.thegreeneconomy.com/apple-shuns-csr-report-at-februarys-shareholder-meeting.
6. Jeff Swartz, "Morality v. Technology? Don't Make Me Ditch My iPhone . . . ," *The Bootmaker's Blog*, May 19, 2011, http://blog.timberland.com/jeff-swartz/morality-v-technology-dont-make-me-ditch-my-iphone.
7. "You Are NOT Allowed to Commit Suicide: Workers in Chinese iPad Factories Forced to Sign Pledges," *Daily Mail*, May 1, 2011, www.dailymail.co.uk/news/article-1382396/Workers-Chinese-Apple-factories-forced-sign-pledges-commit-suicide.html#ixzz1PIflQHzb.
8. David Barboza, "Explosion at Apple Supplier Caused by Dust, China Says," *New York Times,* May 24, 2011, www.nytimes.com/2011/05/25/technology/25foxconn.html.
9. Apple Supplier Responsibility: 2011 Progress Report, Apple web page, http://images.apple.com/supplierresponsibility/pdf/Apple_SR_2011_Progress_Report.pdf.

10. Larry Dignan, "HP Expands Printer Market Share," *ZDNet*, December 6, 2010, available at www.zdnet.com/blog/btl/hp-expands-printer-market-share/42400.
11. Heather Clancy, "HP Discloses Latest Green Policy Progress," *ZDNet,* May 6, 2011, www.zdnet.com/blog/green/hp-discloses-latest-green-policy-progress/17352?tag=content;search-results-river.
12. Tom Raftery, "Have HP's Senior Executives Lost Interest in Sustainability?" *Greenmonk* blog, March 22. 2011, http://greenmonk.net/have-hps-senior-executives-lost-interest-in-sustainability/#ixzz1PNI4u2pw.
13. Ryan Gosling and John Prendergast, "Congo's Conflict Minerals: The Next Blood Diamonds," *Huffington Post,* April 27, 2011, www.huffingtonpost.com/ryan-gosling/congos-conflict-minerals-_b_854023.html.
14. Mark Koenig, "Curbing Congo's Conflict Minerals," *Swords into Ploughshares* (blog), May 8, 2011, http://presbyterian.typepad.com/peacemaking/2011/05/congos-conflict-minerals.html.
15. Bryan Walsh, "Got Yttrium?" *Time Magazine,* March 28, 2011, www.time.com/time/magazine/article/0,9171,2059636,00.html#ixzz1H9AUhG6g.
16. 2010 Critical Materials Strategy Summary. U.S. Department of Energy, www.energy.gov/news/documents/Critical_Materials_Summary.pdf.
17. "New Dawn for Mining at the Seabed," *NewScientist*, June 29, 2011, www.newscientist.com/article/mg21128192.600-new-dawn-for-mining-at-the-seabed.html.
18. Leon Kaye, "Dell Reduced Packaging by 18 Million Pounds since 2008," *TriplePundit,* August 26, 2010, www.triplepundit.com/2010/08/dell-reduced-packaging-by-18-million-pounds-since-2008.
19. "Cisco Packaging Update," www.greenbiz.com/sites/default/files/Cisco_SustainablePackagingUpdate_Feb2010%5B1%5D.pdf.
20. "Sprint Releases Findings of Industry-First Trucost Supply Chain Carbon Report," Trucost web page, March 16, 2011, www.trucost.com/news/114/sprint-releases-findings-of-industry-first-trucost-supply-chain-carbon-report.
21. Todd Woody, "I.B.M. Suppliers Must Track Environmental Data," *New York Times*, April 14, 2010, http://green.blogs.nytimes.com/2010/04/14/ibm-will-require-suppliers-to-track-environmental-data.
22. Joshua Jacobs, "Jim Miller, LGO '93, VP of Worldwide Operations at Google: In the Engine Room of Cloud Computing," MIT LGO web page, June 20, 2011, http://lgo.mit.edu/news_articles/miller-alumni-google/miller-alumni-google.html.
23. "Google's Green PPAs: What, How, and Why," *Google* blog, April 29, 2011, http://static.googleusercontent.com/external_content/untrusted_dlcp/www.google.com/en/us/green/pdfs/renewable-energy.pdf.

24. "Google Buys Oklahoma Wind Power," *Wind Turbine Zone*, April 21, 2011, http://windturbinezone.com/news/google-buys-oklahoma-wind-power.

25. David Worthington, "Google's Clean Energy Investments Nearing $1 Billion," *SmartPlanet,* June 22, 2011, www.smartplanet.com/blog/intelligent-energy/google-8217s-clean-energy-investments-nearing-1-billion/7250.

26. Katie Fehrenbacher, "Google Has Now Invested over $780M in Clean Energy," *GigaOM*, June 22, 2011, http://gigaom.com/cleantech/google-has-now-invested-over-780m-in-clean-energy.

27. "The New Google Hamina Data Center," *LocalFind*, May 25, 2011, www.localfind.biz/the-new-google-hamina-data-center.

28. "Data Center Best Practices," Google web page, www.google.com/corporate/datacenter/best-practices.html.

29. Jeffrey Marlow, "Q&A: Google's Green Energy Czar," *New York Times*, January 7, 2010, http://green.blogs.nytimes.com/2010/01/07/qa-googles-green-energy-czar/.

30. Bilal Zuberi, "Google's Massive Solar Complex and Other Clean-Tech Initiatives," *BZNotes* blog, June 23, 2007, available at http://bznotes.wordpress.com/2007/06/23/googles-massive-solar-complex-and-other-clean-tech-initiatives.

31. "Google Customer Case Study," Bloom Energy, www.bloomenergy.com/customers/customer-story-google.

32. Bob Keefe, "Meet Google's Chief Sustainability Officer (what a cool job!)," DivineCaroline.com, www.divinecaroline.com/22277/44799-meet-google-s-chief-sustainability-officer/2#ixzz1Rw6rkuQx.

Chapter 18

1. Steven Chu, "BP Oil Spill Update," U.S. Department of Energy, August 10, 2010, http://energy.gov/articles/bp-oil-spill-update.

2. "National Highway Traffic Safety Administration, Toyota Unintended Acceleration Investigation," NASA Engineering and Safety Center Technical Assessment Report, January 18, 2011, www.nhtsa.gov/staticfiles/nvs/pdf/NASA_report_execsum.pdf.

3. Joshua Green, "Exclusive: How Steven Chu Used Gamma Rays to Save the Planet," *The Atlantic*, May 13, 2010, www.theatlantic.com/technology/archive/2010/05/exclusive-how-steven-chu-used-gamma-rays-to-save-the-planet/56685.

4. Charles Arthur, "Software Patents 'Gumming Up Innovation,' Warns Chief Google Lawyer," *The Guardian*, July 26, 2011, www.guardian.co.uk/technology/2011/jul/26/google-software-patents-warning.

5. Rafe Needleman, "Google-Motorola: Patents of Mass Destruction," *C/Net*, August 15, 2011, http://news.cnet.com/8301-19882_3-20092679-250/google-motorola-patents-of-mass-destruction/#ixzz1VFPwbYxu.
6. James Surowiecki, "The Regulation Crisis," *The New Yorker,* June 14, 2010, www.newyorker.com/talk/financial/2010/06/14/100614ta_talk_surowiecki#ixzz1SphM5j2A.
7. Mark Cuban, "What Business Is Wall Street in?" *BlogMaverick*, May 9, 2010, http://blogmaverick.com/2010/05/09/what-business-is-wall-street-in.
8. Geoff Brumfiel, "Nuclear Agency Faces Reform Calls," *NatureNews*, April 26, 2011, www.nature.com/news/2011/110427/full/472397a.html.
9. Wil S. Hylton, "What Happened to Air France Flight 447?" *New York Times Magazine,* May 4, 2011, www.nytimes.com/2011/05/08/magazine/mag-08Plane-t.html?_r=3&hp=&pagewanted=all.
10. "MPs Publish Report on Government and IT," UK Parliament, July 28, 2011, www.parliament.uk/business/committees/committees-a-z/commons-select/public-administration-select-committee/news/report-on-government-it-published.

Chapter 19

1. Katherine Rosman, "Eat Your Vegetables, and Don't Forget to Tweet," *Wall Street Journal,* June 16, 2011, http://online.wsj.com/article/SB10001424052702303714704576385652183593900.html.
2. Janet Kornblum, "No Cellphone? No BlackBerry? No e-Mail? No Way? (It's true.)," *USA Today*, January 11, 2007, www.usatoday.com/tech/news/2007-01-11-tech-no_x.htm.
3. Nathan Barry, "100 Things Your Kids May Never Know About," *Wired*, July 22, 2009, www.wired.com/geekdad/2009/07/100-things-your-kids-may-never-know-about.
4. Interview, PBS Frontline, www.pbs.org/wgbh/pages/frontline/kidsonline/etc/notebook.html#ixzz1T3lnLoXC.
5. "Busy NYC Starbucks Block Sockets to Free Up Seats," *Reuters*, August 4, 2011, www.reuters.com/article/2011/08/04/us-coffee-starbucks-idUSTRE7736OY20110804.
6. "Xbox Gamer Dies of Blood Clot after Marathon Session," FoxNews.com, August 1, 2011, www.foxnews.com/health/2011/08/01/xbox-gamer-dies-blood-clot-after-marathon-session.
7. "Healthy Gaming Guide," Xbox web page, www.xbox.com/en-US/Live/HealthyGamingGuide.

8. Jesse Emspak, "Nintendo Downplays 3D's Health Risks," *International Business Times*, January 20, 2011, www.ibtimes.com/articles/103135/20110120/nintendo-downplays-3ds-health-risks.htm.
9. Tara Parker-Pope, "An Ugly Toll of Technology: Impatience and Forgetfulness," *New York Times*, June 6, 2010, www.nytimes.com/2010/06/07/technology/07brainside.html.
10. "reSTART Internet Addiction Recovery Program," press release, November 16, 2010, www.netaddictionrecovery.com/our-mission/press-release.html.
11. Evan Ratliff, "Writer Evan Ratliff Tried to Vanish: Here's What Happened," *Wired,* November 20, 2009, www.wired.com/vanish/2009/11/ff_vanish2/.
12. Social Intelligence web page, www.socialintelligencehr.com/home.
13. Jennifer Preston, "Social Media History Becomes a New Job Hurdle," *New York Times*, July 20, 2011, www.nytimes.com/2011/07/21/technology/social-media-history-becomes-a-new-job-hurdle.html?_r=1&pagewanted=2.
14. Ethisphere, "2011 World's Most Ethical Companies," http://ethisphere.com/wme2011.
15. Don Rickert, "The Ethics Problem in Technology Companies and What Can Be Done About It," *Research, Design and Invention* blog, February 7, 2005, http://donrickert.typepad.com/research_and_design/2005/02/the_ethics_prob.html#more.
16. Vinnie Mirchandani, *The New Polymath* (Hoboken, NJ: John Wiley & Sons, 2010).
17. Matthew Danzico, "Cult of Less: Living Out of a Hard Drive," *BBC News*, August 16, 2010, www.bbc.co.uk/news/world-us-canada-10928032.
18. Smart Product Privacy—2011 update from Mary J. Cronin, author of *Smart Products, Smarter Services* (Cambridge, UK: Cambridge University Press, 2010).

Chapter 20

1. "The Top 500 2011 List," *Internet Retailer*, www.internetretailer.com/top500/list/.
2. Scott Patterson, "The Minds Behind the Meltdown," *Wall Street Journal,* January 22, 2010, http://online.wsj.com/article/SB10001424052748704509704575019032416477138.html.
3. Evan Schuman, "What's the Matter, Wall Street? Amazon's Soaring Profits Not Enough for You? The Attack on IT Investment," *StoreFrontBackTalk*, January 31, 2011, http://storefrontbacktalk.com/social-networks/whats-the-matter-wall-street-amazons-soaring-profits-not-enough-for-you-the-attack-on-it-investment/#ixzz1SqLnvCnx.

4. Jeremiah Wilton, "EC2 Outage Reactions Showcase Widespread Ignorance Regarding the Cloud," *BlueGecko* blog, April 22, 2011, www.bluegecko.net/amazon-web-services/ec2-outage-reactions-showcase-widespread-ignorance-regarding-the-cloud.

Endgame

1. "Overview," Feld Group Institute web page, www.feldgroupinstitute.com.
2. Youtube video, "Thank You Fans from NFL Players," July 21, 2011, www.youtube.com/watch?v=p2E1bOBXAgE&feature=player_embedded.
3. "The Top 500 2011 List," *Internet Retailer*, www.internetretailer.com/top500/list/.
4. Amy Tsao, "Will Wal-Mart Take Over the World?" *Bloomberg Businessweek*, November 27, 2002, www.businessweek.com/bwdaily/dnflash/nov2002/nf20021127_4108.htm.
5. Thomas Wailgum, "How Wal-Mart Lost Its Technology Edge," *CIO*, October 4, 2007, www.cio.com/article/143451/How_Wal_Mart_Lost_Its_Technology_Edge?page=2&taxonomyId=3066.
6. Matthew Boyle and Douglas MacMillan, "Wal-Mart's Rocky Path from Bricks to Clicks," *Bloomberg Businessweek*, July 21, 2011, www.businessweek.com/magazine/walmarts-rocky-path-from-bricks-to-clicks-07212011.html.
7. Ibid.
8. Angel Djambazov, "A Guide to Why Amazon Is Losing the Tax Battle," *GeekWire*, July 3, 2011, www.geekwire.com/2011/guide-amazon-losing-tax-battle.
9. Allison Enright, "Amazon's Problem, Wal-Mart's Opportunity," *Internet Retailer*, March 3, 2011, www.internetretailer.com/2011/03/03/amazons-problem-wal-marts-opportunity.

About the Author

The wide coverage in the book reflects Vinnie Mirchandani's varied career as a technology advisor, industry analyst, entrepreneur, and consultant.

He is president of Deal Architect Inc., a technology advisory firm that helps clients take advantage of disruptive technology trends such as cloud computing and emerging technology innovation areas before they go mainstream. Between this firm and at a previous role at the technology research firm Gartner, Inc., he has helped clients evaluate and negotiate more than $10 billion in technology contracts. The Black Book of Outsourcing has recognized his firm as one of its top advisory boutiques.

He is author of *The New Polymath* (Wiley 2010), a widely praised book—one reviewer called it "an innovation firehose"—which analyzed the coming convergence of infotech, biotech, nanotech, cleantech, and healthtech.

He also writes two well-read technology blogs. New Florence. New Renaissance has focused on innovation for more than six years and has cataloged more than 2,000 innovative technologies, companies, and projects. They cover more than 40 technology categories, from mobile computing to smart products. His other blog, Deal Architect, focuses on economics and waste in technology and was rated by The Industry Standard as one of its favorites.

He spent five years as an analyst at Gartner assisting clients understand trends in enterprise software, systems integration, and outsourcing markets. He spent his early career at Price Waterhouse, first as an accountant and then as a technology consultant. (That division is now part of IBM.) Much of his work at Price Waterhouse was international. There and during his career since, he has traveled to nearly 50 countries.

Mirchandani has been quoted in most major technology and business publications and presents widely on technology enabled innovation. He went to school at Texas Christian University, in Fort Worth, Texas.

Tampa, Florida, is now home. His wife, two teenage kids, and a beagle are frequent "sufferers" of his endless stories of technology innovation and waste.

Index